THE SOUL OF RISK

ISBN 978-1-960378-24-8 (hardcover)
ISBN 978-1-7365373-7-4 (paperback)
ISBN 978-1-7365373-8-1 (eBook)

1st Edition

Cover design by Anna Hall

The Soul *of* Risk

*How a Career in Digital
Risk Can Transform You
into a Modern-Day Hero*

KEN BINGHAM

INTRODUCTION

A Note about Digital Risk as a Career

Since you are reading this book, either you are already on an amazing path toward making a difference (and you need confirmation that you are on the right path) or you've stumbled upon this book as a matter of randomness and luck. Either way, it is worthwhile to consider whether this career/craft in digital risk might be for you. Whatever the case, it's worth it to spend the time necessary to investigate what you will do in this first or next phase of your career. Choosing a career with purpose can be a glorious thing. It becomes your passion, and in a way that ensures you never see your career as *work*. Rather, it becomes one of the things you live for. That passion can take you to the end, or to your next career, or an early exit when you sell the company you founded.

Whatever the case, know this: Others don't have your heart, your dreams, your perspective, or your thoughts.

Those who work with you may have pursued other dreams, working in their lane, and seeing their time like an audition for that something *next*. They may have moved on, created their own consulting company, written a book, become a leader in another company, stepped into another career, but always with a greater strength and conviction from knowing you, a person with clarity and purpose in your life.

You are fortunate to be a part of a workforce era unlike anything we've seen for the last hundred years. The last time we have seen this much opportunity for disruptive change was during the period of industry electrification. But don't just take my word for it. Let's ask one of the elements that's leading to this once-per-century level of change.

Here's what ChatGPT thinks about the situation:

> *The current technological revolution, particularly the rise of the internet, data, and artificial intelligence (AI), indeed represents a transformative period not seen since the advent of electric power distribution. This era is creating unprecedented opportunities for career development and innovation.*
>
> *Firstly, the internet has democratized access to information and communication in a way that is unparalleled in history. It has transformed every aspect of life, from how we communicate to how we work, learn, and entertain ourselves. The ability for instant, inexpensive communication across the globe has fundamentally changed society.*
>
> *Secondly, data has become a critical asset in the modern economy. It informs decision-making,*

drives innovation, and improves efficiency across all sectors. The sheer volume of data generated today, coupled with sophisticated analytics, offers unique opportunities for those skilled in data science and analytics.

Lastly, AI is opening new career opportunities by shifting traditional job requirements and introducing roles that did not exist a generation ago. AI's ability to learn, adapt, and perform complex tasks is creating jobs in AI development, machine learning, robotics, and data analysis. These technologies are reshaping industries and making way for new forms of employment that require new skills and mindsets.

This moment in history, where digital technologies are as transformative as electricity was in the past, indeed presents a once-in-a-lifetime opportunity for individuals to forge new career paths, innovate, and contribute to shaping the future. For those willing to learn and adapt, the potential is boundless.[1]

This book is intended to make you aware that your path, your career, matters even more than you may realize. It's also about the randomness and luck courtesy of the universe,

---

1   Sources:

    pewresearch.org - 3. Fifty-fifty anecdotes: How digital life has been both ...

    wikipedia.org - Data

    linkedin.com - Why AI Will Open Up New Career Opportunities For Young ...

randomness and luck that will make you wildly successful, should you choose a career in digital risk. This era matters, making you part of the most fortunate workforce in the last hundred years. You are in the right spot at the right time!

Most everyone wants a cool and amazing career. Being cool and amazing might last for a moment, but making the best career choice can last for a lifetime.

A New Path:
Discovering Purpose in Digital Risk

In the rapidly evolving job market, finding a career that not only sustains you financially but also fulfills you personally has never been more crucial. The question becomes, how can an unconventional career lead to unprecedented levels of purpose, happiness, and self-worth?

I think I have an answer, and in large part, it's because I've lived it. My story includes pursuing passion over conformity. But you don't know me personally, and in case you don't know of me (apart from my role as the author of this book), then maybe we should consider someone more famous. How about Steve Jobs? The story of the cofounder of Apple exemplifies the importance of following one's passion and intuition against conventional wisdom. Jobs's journey from dropping out of college to building one of the most successful companies in the world underscores the message that following your unique path can lead to tremendous success and personal fulfillment.

While I'm not exactly Steve Jobs, my unique path started seven years ago, when looking for the evidence and the data to show hazardous sites their strengths and weaknesses in risk & safety. I wanted to figure out how to use data as a safeguard. It became an obsession for me. Creating a platform of visuals for sites and organizations to see their risk & safety as more real and more impacting, thereby enabling better daily decision making, became my mission. Most sites and organizations were missing enormous opportunities to harness the data from other similar hazardous sites to both keep themselves safer and save themselves millions.

Given the recent emphasis on using data, AI, and machine learning (ML) to drive decisions and innovation, in the coming pages, I will share how these technologies can be applied to career development and fulfillment, and how, as a new intersection of technology and personal growth, this is *completely* fascinating. AI is extremely important to the risk & safety business at this juncture, and it is very much here to stay. The power of AI is that it enables the people at hazardous sites to make exponentially better decisions, produce higher quality and risk accuracy, while ensuring that all risks (even those risks borne by other sites) are considered in their entirety. Applying these technologies to safety and risk is now a lasting path, one that's sustainable far into the future. AI and ML and semantic search make this choice a lifelong career, with purpose and with never-ending room for learning and building.

In this way, part of the purpose of this book is to present you with a career path that has real, genuine, and lasting purpose. Having successfully navigated this path, this author hopes to capture your emotional engagement, convincing you to become obsessed about doing good and solving problems that are not immediately seen as problems, even though they are always costly and can have an enormous impact on society.

The story in the pages to come is also about Risk Alive, a company built from scratch and guided with a vision to learn from its successes and failures, while building assets and data, using AI/ML to save lives, and benefitting from a lot of that same kind of luck and randomness provided by the universe. In the pages to come, I will share tips, guidance, and learnings from the past seven years, and in the process, will provide confidence in (and, I hope, excitement about) this new and emerging marketplace.

Be Stingy with Your Time

Time is nonrefundable, and how you measure time does not change. It's only your perspective of time that changes. So be stingy with your time. Not having a career and a passion is like not respecting your time, thinking there is always tomorrow. Not true. Living this way eats up our strength, your health, your focus, and your life.

Conversely, you are *instantly* significant when others see that you have a path. They secretly want to be just like you but will never say it. If they are lost and without purpose, be curious enough to ask them "Why are you here?" And be in a position to connect with them, always wondering why you were brought together. What was the universe thinking anyway? It's your job to find out. This could be randomness and luck in your favor.

These days, so many leaders and industries are loudly saying, "Come over here and learn about this. This is the future. The money is great. This is perfect for you." It's exciting, cool, and interesting to be recruited in this way. Yet, if you read my story, these are the wrong reasons to change paths or start a new path. Sure, be curious, listen, but don't be knocked off your path. But be stingy with your time. Don't assume their assumptions or opinions are true. Assumptions are seldom true. Do not get knocked off your path.

Money Matters:
Treat it Well and It Will Serve You Well

My story also highlights some of the financial peaks and valleys that can come from investing yourself in this business. There was a ten-year period where work became so demanding that I missed out on things like grandkids' graduation, family events, and vacations that had to be shortened to attend business events. There were also personal dreams sacrificed, like running a marathon, entering a strongman competition, taking a writing course, learning to sing, and so on.

When I write that you should pursue your passion, be stingy with your time, and recognize that sometimes the career can be demanding, I am in no way suggesting that you should sacrifice in a similar way. Give yourself a time window, with lots of breaks along the way to ensure that you live life rather than letting life live you. As Eckhart Tolle once wrote, "Life is the dancer, and you are the dance." This suggests a philosophy of embodying the experience of life fully, rather than trying to lead or control it from an external standpoint. Tolle emphasizes the presence and living in the moment as pathways to personal and spiritual fulfillment.

So let's apply this perspective to making the right career choice. Imagine how good your life could be if your career returned the money to fit your balance of all your goals! Treat money well and it will serve you well.

Set goals. Chop up your dreams into sections, with early exits, and resets, to ensure that you *are* the dance. This will give you the confidence you need to circumvent roadblocks, money, and people who might stand in your way, all while enjoying life, family, and the dance itself.

Keep to the old farming ethics of working hard and staying focused, and the money will take care of itself. But also keep in mind that working hard does not always mean success. It takes randomness and luck to be amazingly successful, rich beyond your dreams, and not working for money your whole life. In the end, financial stability is important, but it should not come at the cost of missing life's most precious moments or neglecting personal growth and happiness.

Is It a Need, a Nice-to-Have, or Just a Want-to-Have?

As a mathematician at NASA, Katherine Johnson's ground-breaking work was critical to the success of the first U.S. manned spaceflights. Her story, popularized by the film "Hidden Figures," exemplifies the impact of pursuing a career driven by need, in the face of societal and professional obstacles. Johnson's dedication to her craft and her pivotal role in space exploration highlight the importance of perseverance and belief in one's contributions to advancing a need, advancing human knowledge, and advancing capability.

The read that you are embarking on—a book about a craft and an unconventional career choice—is not about a trend or a fad. My hope is that it will help you find purpose and passion for solving an industry need. If you pursue a craft, service, or business that is valuable, choosing a career seen as a need rather than a nice-to-have, the rewards will be more considerably than you can imagine right now. Not only can you join an exciting new paradigm called digital risk, but you can use your craft and your skills to make the world a better and safer place for everyone.

CHAPTER 1

On the Future of Risk

WE HAVE REACHED a tipping point. All industries that deal with hazardous materials or processes are in dire need of new thinking and new people. I like to think of these new people as heroes, the kinds of people who are ready and equipped to work with new technology and data, people with the passion and insight necessary for keeping people safe, people who are interested in using data as a new kind of safeguard for the future.

I don't use the term "hero" lightly either. Think of it this way: Any person in a risk & safety role who helps their site escape an unsafe situation has (in no small part!) saved their employers from lost efficiency, lost money, and potentially bad press. More importantly, this same hero will have likely saved people from injury, saved lives, and helped save the local environment from catastrophe. What could be more heroic than that?

The heroes this industry are looking for are technologically savvy. They are incredibly knowledgeable about how to keep a hazardous site safe. They have a great memory that they can use to reference safety standards, engineering standards, and design guidelines. They have a passion to do the right thing, no matter what.

These days, that last quality is especially important. It can take courage to do the right thing, say the right thing in front of peers, challenge wrong opinions from those who have greater authority. This is an essential component for building a career in risk & safety.

Since you're reading this book, my hope is that you are just such a hero, and that you view this prospect of building new safeguards for the future as an exciting one. Driven by data, there is a bold and thrilling future ahead of us. But the first hill we must clear is to get the word out about what needs to happen, why it needs to happen, and how we, as the future of this industry, are going to *make it happen.*

So, as the old adage goes, let's start with *why.*

Over the years, in part because of flawed processes, in part because of legal and insurance concerns, and in part because of general complacency, managers and owners of hazardous operations seem to have gotten comfortable with not quite doing everything they can to *ensure* safety and instead are just trying to make their sites are "safe *enough.*"

The question of "safe enough?" no longer suffices. The time has come for these managers and owners to start thinking about (and worrying about) all those unanswered questions that have the potential to cause disaster. Complacency breeds unease. We need to be asking, "Have we overlooked any hazardous scenarios?" Without a comprehensive understanding of the risks within a facility, all other work activities stand on shaky ground, risking inaccuracy, wasted efforts, and poor financial ROI.

And that's just the cost and efficiency perspective! Consider the real-world impacts of "safe *enough*." A single overlooked hazard can have far-reaching consequences, jolting not just the immediate stakeholders but also the wider public and the world at large. A case in point is the Baltimore Key Bridge incident, where one stray cargo ship collided with a single bridge support, collapsed the bridge into the bay, and utterly shut down one of the most important shipping hubs on the planet. The financial losses were almost immeasurable. The ecological impact is something we may never get a full understanding of. This event underscores the vital need for rigorous and thorough risk assessment in preventing such unforeseen disasters.

As you'll see in the pages to come, I'm a big proponent of the kinds of insights one can glean by having conversations with ChatGPT. When asked about the Baltimore Key Bridge hazardous scenario, here's what ChatGPT had to say:

> *The Baltimore Bridge collapse had significant impacts, including the destruction of the Francis Scott Key Bridge after a cargo ship collided with one of its supports. This incident resulted in vehicles and people plunging into the water below, with hazardous materials being damaged and causing environmental concerns due to the release of toxic substances into the water.*
>
> *In terms of unconsidered hazard scenarios, it appears that the possibility of a large vessel losing*

power and colliding with the bridge, causing such catastrophic damage, may not have been fully accounted for. This type of event is deemed "extremely unusual" by experts, suggesting that the bridge's vulnerability to ship impacts was underestimated. Additionally, the environmental impact of hazardous materials being released into the water due to the collision was likely not fully considered, which poses significant risks to the local ecosystem and public health.[2]

A 5-day risk assessment produces 1600 hazard scenarios & learnings of what can go wrong.

2   cbc.ca - Baltimore bridge collapse was a failure, engineers say
cnn.com - What we know about the Baltimore bridge collapse
apnews.com - What we know about the Baltimore bridge collapse
washingtonpost.com - How the Baltimore Bridge collapse could impact
cbsnews.com - What to know about the Francis Scott Key Bridge collapse
reuters.com - Baltimore bridge collapse: What happened and what is the

If questions like, "Have we overlooked any hazardous scenarios?" remain unanswered, then how exactly are we supposed to trust any of an organization's work activities that are based on "knowing the risk" in a facility? How much work will happen based on a faulty understanding of what risk remains? What percentage of this work can we consider wasted effort and spending?

More importantly to the world at large, it takes just *one* missed hazard scenario to create a massive financial, civil, and ecological disaster. The Key Bridge support hit by a cargo ship certainly impacted all of these. We're still wrapping our heads around just how much money has been lost to the event. Thankfully, it wasn't on quite the ecological-disaster-level as, say, the *Exxon Valdez* or Deepwater Horizon. But make no mistake: the environmental damages are still considerable.

If we aren't asking the tough questions every day, and if we aren't relying on the vast amounts of data very much available to us in this industry, then it's just a matter of time before the next disaster. Risk does not care about time. It doesn't care if it's been five years or five days since a site's last risk assessment. It sees the first day of operations the same as that one fateful day after years of accident-free operation.

Thanks to some long-overdue changes in an industry that has been doing things the same way for about fifty years now, and thanks to some recent advancements in how we manage data and technology in the risk sector, we have finally reached a turning point. What we need now are the *people* to help us use the data as a safeguard that will ensure these catastrophes never happen again. And that's what this book is designed to do, help us answer an incredibly and increasingly important question:

Are *you* the hero we need?

Revolutionizing Risk & Safety

The landscape of risk & safety is undergoing a seismic shift, driven by the innovative perspectives of young professionals. These professionals are charting new career paths and transforming traditional roles across a global network of sites. Merging their novel insights with established knowledge, they transform isolated learning into collective intelligence, and in the process, they're challenging the conventional notion of "safe enough." As members of this growing and critically important industry, our expertise in data, AI, and forward-thinking empowers us to redefine standards and push the boundaries of how safe these hazardous sites, and indeed the world, can be.

The future belongs to the youth, masters of the tools they know and can adeptly use. As we move forward into this bright future, we the "old guard" encourage them to translate our risk assessments, Process Hazards Analysis (PHA), accident reports, and audits into actionable insights.[3]

To be perfectly candid, this book has undergone a number of different iterations. This is in part because risk & safety has changed drastically in the years it has taken to write and

---

3   Note: For a full definition of terms like "PHA," see the glossary of terms in Appendix B of this book.

rewrite it. By this iteration, my goal is to captivate readers by spotlighting pivotal trends that are revolutionizing risk & safety management. These trends are not only creating new career opportunities but are also addressing enduring challenges across a multitude of sites globally. In this data-driven and AI-enhanced age, we as an industry are seizing a unique opportunity to inspire young, passionate innovators, offering them meaningful careers while bolstering safety for operators and owners, industries, communities, and the world at large. As we navigate this evolving landscape, we are securing data and anticipating the next blend of career and technology that will define the future of risk & safety. There has never been a better time to consider a career in this field.

An Unconventional Career— One that Could Change the World

In the traditional realm of Risk Services, the local expert, akin to the neighborhood plumber, was the go-to for any immediate issues on any site managing hazardous materials or processes. However, the post-COVID era has ushered in a remote workforce, still insufficient against the surge of innovative learning from the younger generation. This period marks the transition from isolated expertise, where a single recent PHA template sufficed, to a dynamic landscape where unconventional methods are paramount.

The past's reliance on static, unformatted tools unsuitable for AI and semantic analysis is giving way to new paradigms. Standard-issue corporate tools that look more like enhanced Excel sheets (tools like PHApro) are suddenly serving as obstacles to the evolution of work processes, data-driven analysis, and AI-enhanced precision. Today's approach to PHA involves teams of up to six members, armed with extensive risk datasets, AI, machine learning tools, and automation, and these efforts are displacing thousands of rudimentary Excel files with insightful visuals and more useful, actionable, dynamic data.

The essence of modern risk assessment lies in the synergy of youth, AI, and data. This combination forms the core of transformative work, depicted in the accompanying visuals, where seasoned advisors outline the periphery, while young professionals drive the innovation within. This integration of data, experience, technology, and diverse perspectives illuminates the meaning behind vast arrays of data, providing clarity and understanding that transcend traditional methods.

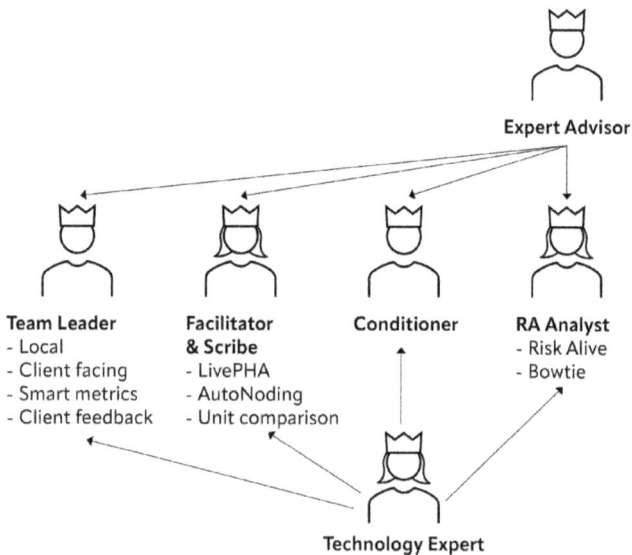

Expert Advisor

Team Leader
- Local
- Client facing
- Smart metrics
- Client feedback

Facilitator & Scribe
- LivePHA
- AutoNoding
- Unit comparison

Conditioner

RA Analyst
- Risk Alive
- Bowtie

Technology Expert

Unconventional Work Process & Crafts is PHAs

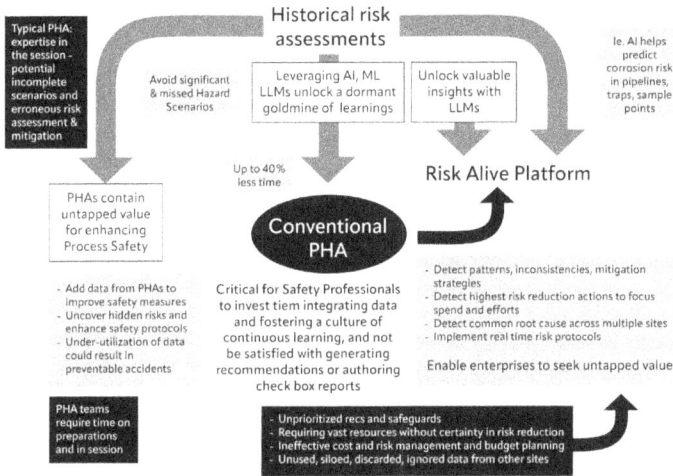

The integration of experience, data, and technology is transforming traditional PHA methodologies, moving beyond mere compliance to enable *comprehensive* risk management. Now, results and insights are accessible across disciplines on a unified platform, allowing site managers to address concerns about hazardous processes with tools like semantic search, unit comparisons, safeguard ranking, and prioritizing recommendations across assets.

In the early days, the goal of leveraging data was a challenge. The potential of untapped learnings was often overlooked, obscured by a lack of understanding. A recent engagement with a growing energy company illustrates this point. They reached out to the author's company, Risk Alive, to explore the hidden value in their data and risk assessments. Despite our efforts to illuminate some overlooked hazardous scenarios—more accurately: overlooked by them but identified by us—the company felt confident in their existing operations team's ability to manage risks at one of their major refineries.

However, within a year, the site experienced a significant incident, resulting in fires and explosions that affected nearby communities. The hazardous scenarios we had previously identified were not acknowledged as contributing factors, as legal and insurance proceedings quickly overshadowed open discussions. The energy company, now focused on insurance claims, lost the opportunity to pursue accountability. The financial and environmental toll was catastrophic. These days, it's an open question about whether the company will even survive.

This incident underscores the importance of comprehensive data analysis, along with the critical need for integrating expansive risk data into safety management practices—using data as a safeguard. Had the company properly utilized historical risk data from similar global sites, they might have averted the disaster.

What are We Up Against?

Risk Alive and I have been driven by passion over conformity, and by our mission of renewing quality and precision in an industry gone stale. We haven't done these things for the comfort and benefit of the largest compliance insurers and services in the world. If Risk Alive had followed this norm, nothing new would have happened. The largest insurers and compliance companies would still be the largest.

The norm would be doing the same conventional things, the same as the largest compliance companies in the world, building similar tools and products (e.g. Excel worksheets, PHApro, and similar products).

It's also important to understand that the primary driver of these unconventional careers is the continuum of significant incidents, and that process incidents, big and small, happen weekly around the world. And each one has the potential to utterly change the world.

Is the Key Bridge disaster not enough to convince you? Let's consult ChatGPT about some of the biggest hazardous site disasters over the past forty years or so...

- Bhopal Gas Tragedy (1984): Following the world's worst industrial disaster, significant changes were made in industrial safety regulations globally. The tragedy led to the establishment of the Environment Protection Act (1986) in India, aiming to prevent such incidents. Efforts by activists and survivors brought attention to corporate responsibility and environmental justice.
- Piper Alpha Disaster (1988): The destruction of the Piper Alpha oil platform in the North Sea led to the Cullen Report, which recommended major changes in the regulation of safety in the North Sea oil industry. Lord Cullen's inquiry transformed safety protocols and led to the offshore industry adopting a safety case approach.
- Deepwater Horizon Oil Spill (Macondo Well, 2010): This disaster had devastating

environmental and corporate impacts. It released millions of barrels of oil into the Gulf of Mexico, causing extensive damage to marine and coastal ecosystems. Thousands of birds, mammals, and sea turtles perished, and critical habitats were severely affected. The spill also resulted in long-term health effects on local wildlife and disrupted fishing and tourism industries. For BP, the disaster led to substantial financial losses, including billions in fines and compensation, and severely damaged its reputation, resulting in a significant decline in its stock value and market trust. Further, it prompted reforms in offshore drilling practices. The establishment of the Bureau of Safety and Environmental Enforcement (BSEE) in the U.S. aimed to enforce safety and environmental protection in offshore drilling operations.[4]

- Mexico City Earthquake (1985): The devastating earthquake led to the creation of stricter building codes and the establishment of the National Center for Disaster Prevention (CENAPRED) in Mexico, focusing on improving earthquake preparedness and reducing vulnerabilities.
- BP Texas City Refinery Explosion (2005): This

4   britannica.com - Deepwater Horizon oil spill - Environmental Impact, Cleanup

natlenvtrainers.com - Environmental Impact of the Deepwater Horizon Oil Spill

ncbi.nlm.nih.gov - The impact of the Deepwater Horizon accident on BP's reputation and stock market returns.

incident resulted in significant changes within BP's safety culture and operations. It also led to broader industry-wide reflections on process safety management practices, influencing regulations and operational practices in the refining sector globally.

- Phillips 66 Houston Chemical Complex Explosion (1989): The disaster at the Houston Chemical Complex prompted reforms in safety standards for chemical plants, including the adoption of more rigorous safety management systems and emergency response strategies. The Occupational Safety and Health Administration (OSHA) revised its process safety management regulations to prevent such incidents.

Nearly 250,000 assets operate with hazards, and nearly all have legal requirements to review the risk related to every significant change and incident every five years. Every *five* years? Could that possibly be good enough? How can this still be the only thing happening in a world of real-time, instant-access data? Even with this kind of risk identification going on, the findings and learnings are siloed, unshared, not quality checked, and not seen as data to be learned from or analyzed. The acts of learning from and analyzing the data are the primary drivers of the new craft this author is hoping to compel you to join, along with all the new careers discussed in this book.

What This Book Isn't

While this book draws from seven years of data experience and twenty-eight years of Risk & Safety experience in numerous industries, it is in no way meant to serve as a self-aggrandizing business memoir. And it tries hard to stay away from opinion and "everyone else has been wrong." It understands that there is nothing to be gained by leaving others feeling like they may be falling behind.

When you have the data—when you have access to intelligence that others don't—it's easy to fall into the trap of thinking you're somehow better, smarter, or more deserving. In fact, when you have this data, the thinking should be more about how to improve people's understanding. In this case, it's a matter of showing a whole industry that they have been doing something fantastic and they don't even realize it. For years, they've been producing information and reports, building up safeguards and risk programs, and gathering evidence of what it takes to do the right thing. They simply don't see it as shareable data . . . yet. But when they do, *look out*.

So this book isn't a "hey, look at me, a guy with a brilliant idea." Rather, this book is intended to help everyone—*everyone*—improve upon how they manage Risk & Safety. There is a reality staring us in the face, one where everyone can leverage a total understanding of Risk & Safety with far greater clarity than ever in all our history. Like PCs, like Google, like smartphones, like ChatGPT, this technology creates a new horizon, one that can leverage existing, siloed understandings into an open sharing model, and for the benefit of every site worldwide that wants to be safer.

A New Frontier

Hundreds of sites worldwide are transitioning from isolated compliance to a model of collaborative learning and knowledge sharing. This shift, underpinned by a commitment to confidentiality and mutual data access, is fostering a global community of risk & safety practitioners. In this dynamic environment, individuals engaged in PHA sessions can tap into a vast reservoir of collective experience, swiftly sourcing answers to pressing questions. For instance, a query about "loss of flow to a crude unit" can be resolved considerably more expeditiously, drawing from an extensive database of 55 PHAs and 19,000 hazard scenarios. This capability is powered by Semantic Search and AI technologies, enabling users to efficiently navigate and extract relevant insights not just from years-old local data, but from the accumulated wisdom of the entire industry (and indeed, the world).

Where before, we were operating on what amounted to educated guesswork, now we have the power of a database containing the sum of all risk knowledge. We have come to an exciting frontier akin to the introduction of broadband internet. What was once a vast amount of siloed information is now fully connected and ready to be made accessible for the benefit of all.

Score	Matched Sentence	Consequence	Cause	Residual Severity	Residual Likelihood	Residual Risk Ranking
1.000	Loss of flow to crude heater	May plug outlet nozzle. Loss of flow to crude heaters. Overheat tube, LOC of crude in firebox. Burn heater down.	Solids accumulate in Crude Degasser, 10-1207	Severity: 4	Likelihood: L2	Risk Ranking: Medium
0.982	Loss of flow through crude heater	Loss of flow through crude heater. Overheat tube, LOC of crude in firebox, Burn heater down.	Block valve closed on outlet of #2 Crude Heater	Severity: 4	Likelihood: L2	Risk Ranking: Medium
0.971	Loss of flow to crude heater	May plug outlet nozzle. Loss of flow to crude heaters. Overheat tube, LOC of crude in firebox. Burn heater down.	Solids accumulate in Crude Degasser, 10-1207	Severity: 4	Likelihood: L2	Risk Ranking: Medium
0.971	Loss of flow to crude heaters	Loss of flow to crue heaters. Overheat tube, LOC of crude in firebox. Burn heater down	Liquid outlet nozzle becomes plugged in Crude Degasser.	Severity: 4	Likelihood: L2	Risk Ranking: Medium
0.942	Loss of flow to crude heater pass	Loss of flow to crude heater pass. Overheat tube, LOC of crude in firebox. Burn heater down.	FV-318 or FV-319 fails closed.	Severity: 4	Likelihood: L2	Risk Ranking: Medium

The above image illustrates how a hazardous site—be it a refinery, nitrogen facility, LNG terminal, or pulp and paper plant—can transform its routine reports (PHA, MOC, LOPA, Near Miss, Accident, Audit) into actionable data. Through analytics and visualizations, it becomes possible to distill thousands of Excel sheets into clear decision-making metrics.

Reflecting on the early days of this new frontier, around 2015, the conversion of analytical data into visuals proved crucial. For example, an owner-operator facing the loss of a site license due to excessive flaring or containment breaches could leverage data analytics to demonstrate thorough investigation and remediation efforts, thereby affirming their due diligence and regaining the operating license.

The visuals depicted stem from initial analytics that identified key safeguards. An ensuing audit revealed that over 30 percent of these safeguards were unaccounted for, lacking verifiable evidence of functionality and reliability from tests or records. These safeguards encompassed procedures, inspections, operator rounds (PRO), basic process control system alarms and trips (BPCS), alarm responses (H&S), and occupancy monitoring (OCC).

Following a Safe Operating System (SOS) audit, which confirmed the unreliability of 30 percent of safeguards, corrective actions were implemented to reinstate operational confidence. Subsequently, a post-SOS assessment verified risk mitigation, leading to the reinstatement of the operating license.

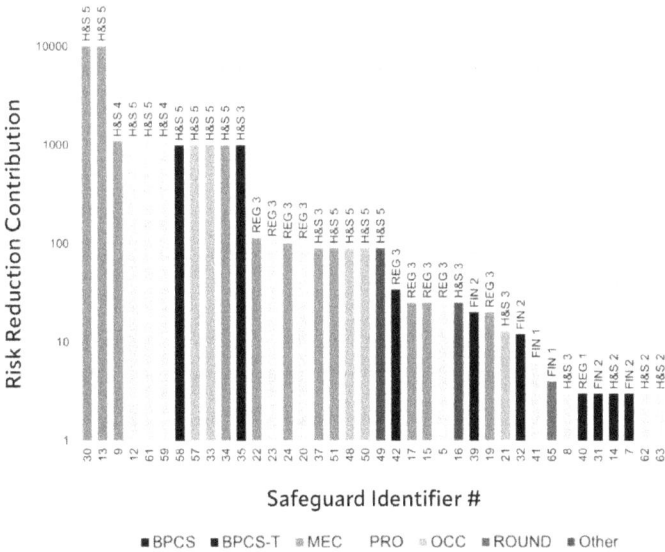

BPCS   BPCS-T   MEC   PRO   OCC   ROUND   Other

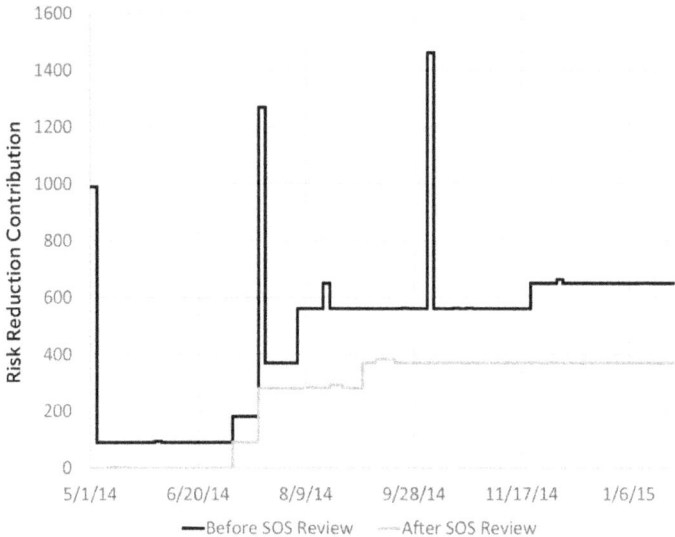

The visuals presented above replace the ambiguity of interpreting thousands of PHA Excel sheets. Disregard the site pricing; the focus here is on the comprehensive and succinct visuals that elucidate the data derived from PHA processes. This material primarily serves those engaged in Process Safety Management (PSM) or risk mitigation activities.

This conventional method's limitations highlight the necessity for a more comprehensive approach to risk assessment, leveraging external data to ensure a more complete risk profile. While longstanding assumptions expedite processes, their accuracy and reliability are often questionable, underscoring the value of a more data-inclusive strategy in contemporary risk management.

Critical Top 3
Communicate the most critical causes, safeguards, and recommendations
$2,990

Report Card
Compare your facility to others using 6 key metrics
$4,290

Recommendation Sequencer
Optimize and reduce spending on recommendations
$7,800

Hazardous Scenario Viewer
Searchable database of hazardous Scenario Bowties
$4,290

Facility Unit Comparison
Understand how your facilities compare and identify missing threats
$7,800

Safeguard Ranking
Identify the most critical safeguards and prioritize for audits or rationalization
$4,290

Moments Matter; This Read Matters

Will this book make a difference in helping you find your career? Whenever I ask that question, I reflect on my own career. I've been at this a long time, and I can tell you that the rewards are incredible. One of the greatest rewards, for me at least, has been the chance to build value every time a service is delivered, over and above being paid for my time. What do I mean? I mean that in addition to getting paid for services, Risk Alive has extracted the learnings from the data collected during those services, and now we're using it to build new value, repurpose it, share it, and resell it. It's why this book is being written. It's the career that this book suggests is worthwhile. This read can lead to a lifetime career change.

Seven years ago, a colleague suggested the thought, implying that the data being collected during our risk service was worth more than the hour of services we were billing. It was the crack-the-nut business we had been searching for. We never looked back. It changed how we served our clients and told our story of value. It changed what we built as a product. It changed how we went to market. It changed one of our basic services forever, conducting Risk Assessments, Risk Identification, and PHAs. It changed our perspective on what we could learn and extract. It became an obsession that continues to this day.

My story doesn't end there. We continue to build new careers, crafts, and tools to fulfill our new service. We continue to re-organize ourselves to increase our efficiency and effectiveness. We continue collecting data worth more than the hours we billed. It turns out our work has many crafts. We leave every client with all our learning, a platform of Risk intelligence to use for the life of any asset.

Our company, Risk Alive, built a new technology using AI, young talent, and changing weeks of work to five-year terms. It was monthly, recurring work analyzing learnings on many assets, turning a one-asset job to an all-in assets job. This is what inspired this book: the call to share the opportunity of nonconventional careers with the world. It's an era to shine for our new career seekers, a once-in-a-hundred-year kind of opportunity.

So congratulations! In exchange for only a few minutes out of your life, you might find yourself in a very unconventional (and extremely rewarding) career. Thanks for your interest.

CHAPTER 2

An Industry in Need

CONSIDER THIS: the world spends an estimated $34 billion—that's $134,000 per site—to conduct risk assessments and then complete all the non-risk-reducing recommendations, just so they can be compliant and legal. At the same time, owner-operators find themselves less profitable whenever they suffer unsafe days and reputation-damaging incidents. These incidents don't even have to happen in their own backyard! Blame for whole industries can be leveled by a single Tweeter, or by one TikTok paintbrush.

Not everyone is open to being humbled by what the data says. Yet there is an emergence of Risk practitioners who are considering saying no to the traditional Risk approach, and no to anyone saying, "You don't need the data," or "We have no time to gather the data," or "We aren't aware of where the data is."

Without the data, many Risk consultants continue their work. Perhaps they don't have to live with the risk like the primary benefactors of safeguards do. This group should be unsurprised when they're met with the same results in the same facilities, repeating the same risk assessment again and again.

The primary benefactors—the owner-operators—should be wondering why we need more risk assessments if we're just going to see the same results. To them, it looks like a poor use of resources, people, and money.

The problem here is multifaceted, but one of the biggest contributors to the problem is related to the workforce. The Risk & Safety industry is in dire need of young people, software engineers, biochemical engineers, environmental researchers, and all-around problem-solvers, but there are just so many other industries to distract talented people from those backgrounds. This leaves the old guard to apply old-guard solutions. And we're all just hanging around doing the same things repeatedly while expecting different results, growing older, approaching retirement. Or maybe we're thinking about heading to greener pastures; maybe trying our hand at software engineering, finding something else that isn't so frustrating to work on.

The solution to this part of the problem is obvious: we as an industry need to compel new graduates looking for a purpose—looking for the opportunity to do some real good in the world—along with people in their thirties who might be looking for a second career where they can apply their own experience in a more meaningful way; and also the entrepreneurs, consultants, and business owners, already in the business who might want to revolutionize what they do and help the world be safer. We need all of these people, and fortunately, the time has never been better to accept them into the fold.

I hope that this book will leave you inspired enough to consider a career in an industry that is stuck and slow to change, one where current change happens almost exclusively after significant and catastrophic events. This is where a first-career person introducing data, science, and apps to the Risk industry can make a huge difference. Being at the forefront of this change will be so significant and so refreshing—maybe even cool and exciting. And if you can get the old guard behind you, then you'll help make this digital Risk transformation even more magical.

This book is also for second-career people and entrepreneurs familiar with the Pacific-Ocean-sized data lakes just waiting to be tapped. The same people are excited about finding the data (especially the Process Hazard Analysis [PHA] data) on the shelves of 250,000 sites worldwide, data just waiting for someone to see it not just as site-specific information, but as cool and exciting learnings whose numbers are as expansive as the stars in the universe. You can then say to the old guard (and to anyone else willing to listen), "This is how you can learn about Risk quickly, make better decisions, work more efficiently, and increase access beyond experts and throughout many different parts of your organization."

You can be one of the revolutionaries who rescues the Risk Industry from the insurers, the guarded old ways, and the traditional, non-curious experts. You can show them all that Risk can be something cool and engaging. You can show them how learning can start all over again and not be ignored any longer. Just as importantly (and convincingly), you can show them how it's too expensive to ignore the data any longer. Just bringing your perspective and learning techniques can help this industry see the light. It can *rescue* industries, *prevent* them from hurting themselves even more than they already do, and *reduce* their spending on safeguards and recommendations that don't actually reduce risk.

And this is only the start.

There is no dark side in this book. My intention is to promote the digital risk industry in a way that ensures everyone will win, no harm is caused to any, and that all sites, employers, and the world as a whole will benefit considerably.

There are many insights about the industry in these pages, but at its heart, this book is about getting an industry that is still operating on twentieth-century strategies into the twenty-first-century mindset of using data as a safeguard. If we're going to achieve that end, then we must recruit great people like you into careers that will help deliver the safeguard to the primary beneficiaries (the owner-operators) and then to the secondary beneficiaries (insurers, regulators, and manufacturers). What we're looking to do here is attract first-career people, owner-operators, and entrepreneurs who see data as an asset. We're also very much open to second-career people who desire to serve a bigger purpose: to make the world safer.

What Say You, ChatGPT?

This book is going to bend some of the rules by creating fictional characters to serve as stand-ins for the kinds of people we're trying to recruit to the industry. It's also going to lean a little on AI to answer some big questions that not enough people in this industry are asking. But first, let me ask you a question:

Are you looking for a role that is both challenging and rewarding? Do you want to make a real difference in the lives of your fellow workers and the community? Then reducing unsafe days in hazardous operating facilities is the job for you! Work with world-renowned safety experts to develop and implement safety protocols, evaluate potential risks, and use cutting-edge technologies to create a safer work environment. Join the fight to ensure our workers—and our communities—never experience an unsafe day at work again!

The need for people to join this fight is enormous. After all, according to ChatGPT,[5] I am working with the world's second most dangerous facility type. Here's the list the AI put together:

---

5   There are, of course, caveats when referencing ChatGPT (which I admittedly do several times in the coming pages). Here's the primary one: the language model based on machine learning taps into what people write and publish on the internet. In other words, it's not coming up with these ideas on its own; it's synthesizing the ideas of a given collective of people. This can of course lead to some dangers if we believe ChatGPT without first inspecting the data sets (which could be biased or even manipulated in some cases).

In our example above, consider that ChatGPT's apparent perception that nuclear power plants are the most dangerous facility type will look quite a bit different if you're in, say, China when you ask it the same question. Likewise, it might look different in Russia, India, France, and so on. While ChatGPT is a useful tool, it must be recognized as a tool that simply expresses something akin to the consensus based on available published information.

1. Nuclear power plants
2. Oil refineries
3. Chemical processing plants
4. Natural gas plants
5. Hydroelectric dams
6. Offshore drilling platforms
7. Deep sea mining facilities
8. High voltage electrical substations
9. Rocket launch facilities
10. Biological/pathogen containment labs

I'm always one to embrace new technologies, so once I got started down the ChatGPT rabbit hole, I found myself asking it all kinds of questions. What I learned could not have confirmed my hypothesis more completely: 1) that hazardous sites are facing more unsafe days than ever before, 2) there is a tremendous need for new lines of thinking on Risk & Safety as a result, and 3) if we want to drive the necessary changes, then we also need a huge number of forward-thinking people to join the cause.

ChatGPT defined the risk of an unsafe day in these facilities as ranging from the accidental release of hazardous or nuclear materials to explosions, fires, or other catastrophic events that may result in serious injury or death to personnel or damage to equipment and infrastructure. In addition, facilities such as nuclear power plants may present the risk of radiation exposure to personnel and the surrounding environment.

Interestingly, ChatGPT couldn't estimate the number of unsafe days and near misses per year for all hazardous operating facilities. It just isn't possible to accurately make such an estimate, as the number of such facilities varies by location, and the type and severity of incidents can vary greatly depending on each individual facility's safety protocols, engineering precautions, and operating procedures. Unsafe days reported by social, regulatory, and interest groups each year are also difficult to compile, as different organizations collect and report data in different ways, and also, members of these groups may not always report all incidents, which means the number reported may not accurately represent the actual number of unsafe days.

So, how big is the problem of unsafe days in these facilities? Is it getting worse? Are we seeing more and more reports of unsafe days?

According to the AI, the problem of unsafe days in hazardous operating facilities is a major issue. It appears to be getting worse, as more reports of unsafe days come in and safety protocols are not strictly enforced. The US Department of Labor statistics reported in 2018 that there were approximately 198,300 reported worker injuries and illnesses that resulted in days away from work, job transfer, or work restriction. Additionally, since 2016, the US Bureau of Labor Statistics has also reported an increase in fatalities in the private mining, quarrying, and oil and gas extraction industries.

The AI suggested that the best way to reduce the risk of unsafe days in these facilities is through a combination of rigorous safety planning, inspections, proper training and drills, strict enforcement of safety protocols, use of appropriate safety equipment, and having redundant systems to help mitigate catastrophic events. Additionally, companies should strive to create a culture of safety, with employees feeling comfortable to report potential safety hazards and deliberately question potential unsafe operations.

It might not sound cool, interesting, and exciting, but fortunately, the chatbot also had something else to say: if all process facilities shared their risks and considerations of potentially hazardous events, threats, and impacts in an open platform, it could reduce the risk of unsafe days. Having a central repository of information that all process facilities could draw from would ensure that all facilities are aware of the potential risks and responses to them to ensure that safety protocols and procedures remain up to date and effective. Additionally, it would be an excellent way to educate and train personnel and stay up to date on industry trends and new technologies.

So, ChatGPT, how many more experts would the industry need?

Answer: It is difficult to estimate the exact number of experts in the world today that specialize in reducing unsafe days in hazardous operating facilities. However, the number of experts is likely not enough to address the problem adequately; more industry-specific safety training, education, and research is needed to mitigate the risks and reduce the number of potentially unsafe days in these facilities.

I can only conclude that there is a clear business case for being a leader with a new craft for assessing unsafe days at hazardous operating facilities. Assessing unsafe days provides actionable data on the current state of safety that can be used to identify potential areas of risk and inform decisions to increase safety. Additionally, assessing unsafe days could help attract and create more experts in the field by providing a tangible metric that shows the effectiveness of new safety measures or technologies.

It is my hope that the pages to come will help attract bright people into what is absolutely, unequivocally (and yes, kind of surprisingly) cool, interesting, and exciting work in Risk & Safety.

Okay, fine! I'll leave you with a back-of-the-brochure kind of pitch:

Are you looking for a role that is both challenging and rewarding? Do you want to make a real difference in the lives of your fellow workers and the community? Then reducing unsafe days in hazardous operating facilities is the job for you! Work with world-renowned safety experts to develop and implement safety protocols, evaluate potential risks, and use cutting-edge technologies to create a safer work environment not just at one site or another, but throughout the world. Join the fight to ensure that our workers, and our communities, never experience an unsafe day at work again!

But What's in it for Me?

If the Higher Purpose isn't enough to sway you about the future of Risk & Safety, maybe we should ask how Risk & Safety could benefit *you*...

At the time of this writing, we are witnessing a recession, one that is raising worldwide nervousness about the foundation of our financial systems, and one which is making more than a few people think about whether their chosen career is genuinely recession-proof. If you're one of those people, then consider the compelling notion that digital safety is an incredibly secure career. Data is the future. This is a fact. And as long as you're working to improve data collection, sharing, and effective use in an industry that desperately needs it, you're safe from any recession.

Become an expert in your company or become a consultant working for a digital safety company, and your career will always be trending up. How do I know this? Because there are at least three classes of people or groups who stand to benefit tremendously from the work of people like you who wish to contribute to this industry. The first group is one that I hope you will belong to after reading this book, a group I call "the owners of the problem." These are the people who will be tasked with delivering on the safeguard called data. These are the recent graduates, the first and second career people, the risk consultants, and anyone currently associated with a risk company.

The second group I refer to as the "primary benefactors." They are those who stand to benefit from all the good that can be extracted from the data. These are the owner-operators, the ones who own, run, and/or benefit from hazardous sites. Those who put their reputations at stake every day their site is in operation.

Then there is the third group, which I refer to as "the old guard." They represent the obstacle we, in group one, must overcome. These are the decision-makers who view data collection as an annoying expense, and data sharing as antithesis to a competitive business. Make no mistake; the old guard is going to be facing an extremely limited future.

It's a beautiful thing to work in a renewed industry like the one before us. Only a few years ago, there was no such thing as the digital safety industry. Today, it's estimated to be worth billions of dollars. It's a beautiful thing knowing that this business is on a rocket ride, and that it won't be burning out anytime soon. After all, once you start using and relying on data, there's no going back.

Though some of the above might sound like hyperbole, make no mistake; this book is not based on speculation, but seven years of collecting, ingesting, and learning side by side with primary benefactors as well as those who are tasked with delivering the data that more and more sites are using as safeguards. This book's claims are based on millions of learnings and thousands of risk datasets. They're drawn from what the data and learnings have already told us about what has happened, been considered, and drawn conclusions from. All of this information came directly from someone like you, someone who has worked for or is considering working for a company that operates one of the 250,000 hazardous sites in the world.

I'm Still Not Sure This Career is For Me

If you're a young person starting out, it might be difficult to see why it's time to think about the possibility of making a career out of helping others be safe. As human beings, we were born with an instinct to survive and take the road well-traveled. This book presents and discusses a new highway.

This is part of why an influx of new talent to the field is so important: there are worlds (and people) out there who desperately need you to change their perspectives. The older generation needs help to take their cars off autopilot. They need to be convinced that it's good to be curious, to test their processes (and themselves) every day, to find new ways to measure and validate safety, and to believe that it's not enough just to create a safe day, but to make sure that today is safer than yesterday.

Because here's the truth: almost all facilities that handle hazardous materials or processes do not see all the risks. Most of them have limited access to learnings that could be gleaned from similar sites or sites with similar equipment or processes. If they had access to each other's data on how to keep this equipment and processes running safely, it would make not just the industry, but the whole world, a better place.

We in the Risk & Safety industry need you, and your new perspectives, and your data, and your insights, if you're willing to share them. We need you to help connect all the data from all the different sites operating worldwide, and we need you to help all those individual sites figure out how valuable it can be to possess and effectively use that data. Make no mistake; they cannot do this without you. They're too stuck in the old way of thinking, that as long as their individual facility is safe, everything is fine. Little do they know that there may be other similar facilities out there that have faced unsafe days because of a piece of equipment or process that wasn't properly monitored or maintenance. If they had access to the data, they could see the potential for disaster before it happened.

If you don't believe me, just consider 2022. I know it's difficult to think about such a weird, often terrible year. But let's run through a quick exercise. End of COVID? Not really, and now we also have monkeypox. End of the endless wars? No, there's a new one in Ukraine, Taiwan, maybe India. End of starvation? No. In fact, thanks to a large number of factors threatening supply chains and impacting food-producing countries, it's only getting worse.

In the Risk & Safety business, we're typically surprised by weird things, but 2022 has made many of those weird things more normal than ever. Even as we work to lessen the impact of workplace hazards and unsafe days, many other things are threatening from the outside. Now inflation, and dollar devaluation. Think about the impact on worlds and companies trying to be profitable through tough times and external factors, all without following everything they can learn from the data.

Fortunately, that's not the only perspective we can draw from 2022. That trial-by-fire that was 2022 helped clarify a central message of this book: the world wants and expects to be safe. Whether the cause was war, pandemics, or poor workplace practices, tolerance for unsafe days began to wane. 2022 helped the world evolve. For the first time, the Risk & Safety industry has enough data, enough science, and soon, enough young people like you who understand how to apply this intelligence to something like safety. With your insights, worlds and companies can scale.

Is it Time for a Second Career?

So you've spent years in a first career in another field but are wondering whether Risk & Safety might be the thing you want to do with the rest of your life. You might be the kind of person who sees the rest of the world changing, accelerating, morphing, leaving you wondering if you're missing out somehow. Probably, you have had some exposure to Risk & Safety in your first career, but it wasn't your primary job. You have wondered about becoming involved, and can relate to the reports, information, procedures, and audits around the process units and equipment you've become familiar with at your current job.

You might not realize it yet, but you create useful information (and even data) in many of the things you do each day. Even while you've been working hard and paying for your mortgage, car, and credit card, you've been creating data. Don't believe me? Try running a Google search for a lawnmower, then count how many days thereafter where you see nothing but popup ads for lawnmowers. The data that came from someone of your background trying to buy a lawnmower makes it all so easy! You were just curious about lawnmowers, and now you have so many options to choose from, all generated by a prediction algorithm that knows quite a bit about what you might be looking for to cut your grass.

What if you could leverage the same processes to identify risk, prevent accidents before they happen, and make the world a safer place? If you're considering a second career, and looking for a greater purpose, consider this: there are more than 250,000 chemical sites worldwide that currently produce and handle hazardous chemicals. What if you could help make all those sites safer?

The industry desperately needs people like you, people who can help it catch up with the data and existing technology. The traditional Risk & Safety business has always been slow to change, leaden with admin and work processes. Lately, investors recognize this situation as a chance to reform, putting a new layer of protection on people and the environment, by using the data to be smarter and see risks more clearly.

This de-risking of all industries is conservatively estimated to be worth $75 billion, according to baseline assumptions. What are those assumptions? We start with 250,000 sites worldwide. We assume that each site spends an average of $150,000 annually on Risk Assessments designed to resolve actions and mitigate risks. Each country requires licensing of these sites, who must validate their risks once every five years in order to maintain that license. This gets us to a total of $37.5 billion dollars for each five-year cycle.

Now, let's also include the cost of significant unsafe days, ones that result in a financial loss due to environmental damage, equipment loss, and disability claims of $800,000 every ten years. This increases the total financial worth of the Risk & Safety industry to $75 billion annually.

Wouldn't you want to be a part of something that huge?

All across the world, every site that handles hazardous chemicals is frustrated. "Why should we continue spending on Risk Assessments, Audits, PHAs, and HAZOPs when all it seems to do, at best, is keep our insurance rates the same, or with minimal increase?" The malaise causes these sites to procrastinate on thousands of recommendations. Those sites eventually incur unsafe days, left wondering how they might have avoided them.

Somehow, even after these unsafe days, even after the environment and sometimes people are harmed, and even after the company's reputation and share price plummets, each site approves new spending on old methodologies, only to have accidents repeated in a few years. Excessive spending on unsafe days, ignoring the data, relying on insurers to clean up the mess—this has been normalized in the industry. The spend continues, even as regulations, technology, metrics, and measures are slow to change.

It doesn't have to be this way! You can fix all of this! This is why the timing for an influx of people who think differently is so perfect. Risk & Safety is a new and booming industry undergoing digital transformation and a massive turnover to younger generations of leadership. If we do everything the right way, then in short order, all those sites will be more efficient, profitable, safer, and will experience fewer unsafe days. The only adverse side effect is that it will take a bite out of the insurance industry, as some facilities start to self-insure and invest in the data to understand their risks more clearly.

Starting a second career in Risk & Safety, you quickly learn to inspire others. You get to help owners with their first-time experience with something extraordinary, elegant, and attractive. You get to inspire the young employees of the owners to sign up with five- and ten-year contracts. When they ask for help to provide a five-day PHA or similar risk study, you comply, hoping they are up for something different, not available anywhere else you are aware of. You go in knowing you come loaded with a different way of thinking and doing, and you are prepared and excited to lead them through this, to see their risk differently and more clearly than they have ever seen it before.

CHAPTER 3

Data as the Twenty-First Century Safeguard

DATA IS FAST becoming the most critical safeguard of our time. The data (or at least the data at question in this book) illustrates the intelligence and the *oneness* of what the world thinks about risk in any specific facility that handles hazardous materials or manufactures dangerous products. Bhopal, Pasadena, BP Texas City, and Piper Alpha—all of these names are synonymous with unsafe days that led whole industries to adopt new safety standards. But in each case, the safety had to be preceded by a disastrous event. An unsafe day.

Today, with our world-sized risk database that allows us to share and compare risk elements related to threats, impacts, hazardous events, and safeguards, we have surpassed every safeguard that ever existed prior to this century. How would you calculate the value of such a safeguard? What would you even *call* it?

The safeguard described above could be called a "Global Risk Intelligence Network (GRIN)." It would consist of a centralized database that collects and analyzes risk data from various sources worldwide, including past incidents, current events, and emerging threats.

As to the value of such a safeguard, it could be calculated by assessing the reduction in the frequency and severity of incidents and accidents in facilities handling hazardous materials or manufacturing dangerous products, as well as the increase in the efficiency and effectiveness of risk management strategies. Additionally, the GRIN could provide valuable insights and recommendations for improving safety standards and regulations, ultimately saving lives and reducing economic losses on an incredible scale.

A Quick Conversation with ChatGPT: How Would Steve Jobs Make GRIN a Reality?

Question: If the construction of GRIN were an Apple project, how would Steve Jobs inspire his company to build it? What would he say? What would be his infamous three steps to execute GRIN and offer it to the world?

Answer: As an AI language model, I can't be sure what Steve Jobs would say, but here's a possible perspective:

Steve Jobs was a visionary leader who believed in using technology to solve complex problems, and his approach to innovation was driven by his passion, creativity, and attention to detail. Suppose Steve Jobs were alive today and leading Apple. In that case, he might inspire the team to build GRIN by emphasizing the importance of safety and security in all aspects of life, including technology.

Here are three possible steps Steve Jobs might take to execute GRIN and offer it to the world:

1. Start with the user experience:
   Jobs famously believed that technology should be designed around the needs and desires of the end user. He might challenge the team to create a user-friendly platform that makes it easy for anyone to access and use GRIN, regardless of their technical expertise.

2. Focus on simplicity and elegance:
   In keeping with Apple's design philosophy, Steve Jobs might encourage the team to prioritize simplicity and grace in the design of GRIN. He might argue that the platform should be intuitive and easy to use without sacrificing functionality or security.

3. Drive innovation with purpose:
   Jobs might motivate the team to innovate with purpose, emphasizing the importance of using technology to solve real-world problems. He might challenge the team to think creatively about how GRIN could enhance safety and security in various industries and contexts, from manufacturing to transportation to emergency response.

Andrea, Our Main Character, Runs with It

This book will occasionally to bend some of the rules by creating fictional characters to serve as stand-ins for the kinds of people we're trying to recruit to the industry. Our main character, the person most eager to spur change, is a fictional second-career person named Andrea.

So, in the spirit of bending the rules, let's take a moment to see what Andrea thinks about all this...

When she first read the feedback from ChatGPT on how Steve Jobs might design and implement GRIN, Andrea had a blank look on her face. Asking the AI for input on the matter had been her coworker Ross's idea. As the youngest member of Andrea's team, Ross had always been focused on doing the right thing for the business and leveraging all the newest technology to get there.

It wasn't the first time that one of Ross's presentations had taken Andrea by surprise, but it was the first time it left her speechless for quite this long.

"Look, I understand that this might seem a little out there at first," Ross said. "But I've been thinking about how to make what we do seem cool, engaging, and exciting for the rest of the staff."

Though she really needed something substantive to take back to her executive VP, Andrea found herself a little overwhelmed by the proposal. "We already have a lot of work trying to automate and become efficient at running our analytics and ingesting the data accurately."

"Yes, but—"

Andrea held up a hand for pause. "Let's package this up and present it to my exec. But first, let's add a timeline. And let's highlight areas where we can get some early return on investment—anything we can show that recovers the work cost and demonstrates how long this would take without incurring additional costs and people. Otherwise, there's no way something on this scale is going to fly right now."

"On it."

Always the eager one, Ross stood to leave, but Andrea motioned for him to remain a little longer.

"The excellent news is that I think we've already finished step one," she said. "We already have a GRIN-like platform in place. It's constructed more for the practitioner at the moment, so there's some work left to make it worthwhile for a nonpractitioner."

"Not sure I follow."

"Let's think about the maintenance guy or operations guy on the plant floor using the platform. They'll want to benefit from GRIN as well. Anyway, just some added work necessary to meet the Steve Jobs first step."

Ross broke into the kind of contented smile that only comes from having one's boss accept his idea. "What about step two?" he asked excitedly.

"As we're refitting the GRIN platform for a nonprac-titioner, we might consider reducing the exposure of functions," Andrea suggested. "We'll want to keep things visual-oriented, with as little text as possible. More visuals and more graphics. No Excel field presentations of the data, in other words."

"I love it."

"More important than any of that, though," Andrea continued, "is that the users have to be empowered with the choice to take any action. Can we provide users with a multiple-level deliverable to help justify decisions to their boss even if they don't like the analytics results? Can we provide them with a metadata version of their raw data that they can use on their own and into the future? This way, the user can say, 'You don't believe me? Hey, let's check the analytics.' If we can achieve something like that, then my bet is that they'll come back to us with more requests."

"Okay, so what about step three?"

Andrea returned to the text of Ross's presentation. "Can we be more creative and think about others? Who else would benefit from the GRIN platform? We've been thinking mainly about the people working with Risk Identification, PHAs, and HAZOPs. What if we included insurers, auditors, and operators, all making their own craft decisions but using the platform to make significant decisions?"

Ross jotted down the notes. "What would that look like?"

Andrea furrowed her brow in thought. "Insurers are more about loss, so events and equipment are impacted. Auditors are more about checking on procedures, testing, and maintenance, so they're more equipment focused. And operators are more about making significant decisions to keep production going safely, so they're going to respond to information about safeguards, bypasses, and any equipment impacted by a given choice."

Silence hung between the two of them for a time before Andrea broke into a knowing smile.

"Ross, you've sold me. Can I suggest packaging this up so I can present it in five minutes to my VP?"

Once she had the information she needed in hand, Andrea hurriedly exited the project office between the engineering services and analytics teams. Several of the staff in each department noticed the extra bounce in Andrea's step, and they responded with a smile. Ross had been anxious about this meeting, and they knew that any positive result would benefit the company. With budget cuts, people leaving unexpectedly, and the high rate of turnover happening since the dawn of COVID, change was needed to stop the bleeding. It looked that Andrea had become a true believer—that this project might be just what they needed not just to save the company, but to get the excitement back for the group.

An Exciting New Technology

Data used as a safeguard to reduce risk (in the form of something like GRIN) will face the same scrutiny as SIS in the months and years to come. But long after it comes into widespread use, the market thirst for technology that is easy, quick, and efficient for every site in the world to use will be sated. The market, once aware that data used as a safeguard is possible (and indeed, is being deployed even as this book is being written), will be hungry for another quicker, low-cost, and easier-to-use safeguard. The data, having laid dormant for so many years, can now be used holistically as a box to be checked during all phases of development—from concept to pilots, through engineering and design, and into operations and maintenance.

The initial hurdle for most new technology and systems is that you must prove financial viability, but in a world with ubiquitous social connectedness and roaring demand to protect the environment, those who are unaware of the cost savings (and life savings) of this new technology will wake up to the reality quickly.

In the near future, regulators, corporations, and insurers will want evidence to prove low-risk operations with manageable Risk and change. Having data as a safeguard will quickly become the risk training for Artificial Intelligence (AI) that will manage these systems. AI will be fed data from a worldwide database, primarily consisting of PHA data sets. It will be trained to process this data so that it can check for missing threats, safeguards, and impacts and therefore help human users to more accurately manage risk.

If we were born today, we would not know any different, and we would simply accept what's in front of us. But since we have lived a while already, we have biases in how we perceive or want things to be. It's hard to see the future, but we must try because we are not in charge of the future; our kids are. There is one thing that's survivable for the future: the data.

But don't take my word for it. Let's ask an AI direction. ChatGPT, what say you?

Question: In the future, investors will see profitability as only one factor, but can it be safe for people and the environment through the change of people, owners, equipment, and tweaks to the process?

Answer: The future increasingly relies on data and the ability to effectively manage and analyze it. As we continue to collect vast amounts of data from various sources, the need to pass on this data to future generations becomes critical. Data is essential for decision-making, innovation, and progress in multiple fields.

Passing on the data baton to the next generation ensures that data is appropriately stored, catalogued, and accessible to future generations. This includes providing tools and technologies that enable easy access and analysis of the data. Additionally, it involves training future generations to effectively manage and analyze the data to ensure that it remains relevant and valuable.

The importance of passing on the data baton cannot be overstated. Data is critical for scientific research, healthcare, business, government, and other aspects of modern society. Without access to historical data, future generations may struggle to make informed decisions, innovate, or address complex challenges.

Furthermore, data is one of the few things that can survive over long periods. Unlike physical objects, data can be stored and accessed indefinitely, making it an invaluable resource for future generations.

In summary, passing on the data baton is crucial for the future of society. It ensures that data remains relevant and useful for decision-making, innovation, and progress and provides a valuable resource for future generations.

CHAPTER 4

First Careers

A FEW WEEKS passed before Joe received a call from Andrea, who wanted to know if he could participate in a new corporate initiative using digital transformation around safety. Joe was a former operator at our organization, and at my urging, had been dabbling at becoming a Risk & Safety specialist. Given his experience as an operator working on site with this new access to all Risk & Safety information, Andrea thought Joe could contribute to the success of our new corporate initiative. She was curious about what the cost of access for all operators might be. When Joe told her that the data and learnings could be shared across the organization with a small, Netflix-type charge for access on all process units, Andrea was thrilled.

"That means no license fees or training would be required," she said.

"Exactly," Joe said. "We could share all risk information based purely on tag number and/or equipment name. Then, anyone who needs the data could look it up by entering those details. It would be a searchable database of knowledge."

This was something new for us, and the excitement quickly spread. Andrea burst into my office and explained the breakthrough. We discussed how, on previous tools, licenses were purchased just for those warranted to gain access to the information. But with this new system, everyone had access, no restrictions, and minimal training, from the night guard to the CEO. This was exciting news.

As I walked away, assuming the discussion was over, Andrea asked me if I could also help her team plan for next year's turnrounds and audits, not only for Joe's site but for all the sites under her jurisdiction. I smiled and tried not to be smug about my comments because I knew that the health checks I hadn't shared yet could quickly help us prioritize the work for upcoming turnarounds and audits.

"I look forward to helping," I said simply. "I just need access to the recent and older Risk Assessments for all these sites. I can then come prepared to highlight our priorities and the results at our next meeting. It should help us understand the criticality and priority of each tag number of equipment and safeguards from a safety perspective."

The Doors of Learning Aren't Always So Wide Open

Derek was a key member of the operations team of the Cracker Cat Unit at Acme Refinery for over ten years. By that point, he knew every valve and every piece of equipment in his unit. When an alarm went off, the folks in the control room would reach him on the radio or his phone, and he could tell them exactly what they needed to do to prevent a bad day from happening. This unit was like home to Derek, one whose every corner he was familiar with.

One day, he got a call from the Area Operations Manager. The most experienced person in the Poly Cat Unit had taken up a different role elsewhere and departed. There had been no knowledge handoff, no training for others on the team, essentially nothing to fill the experience vacuum the experienced person had left behind. All this, and they had an Alarm Rationalization Study scheduled for the next week. This meant they had to catalogue every alarm in the unit and document actions to take, time to respond, and more for *every alarm in that unit*, all without the knowledge and oversight of their most experienced member.

"So we're wondering," the Area Operations Manager said to Derek sheepishly, "given your experience, maybe you could support some of our lesser-experienced operations team members with the study..."

Derek held the phone away from his face and sighed. If this had been his own unit, he wouldn't have hesitated to say yes. Some of the Poly Cat operations were familiar, but at the end of the day, he didn't know that unit's pulse like he did the Cracker Cat's. He didn't want to say yes without being completely confident that he could quickly get up to speed, because even a small miss on his end, and someday an alarm could go off, and nobody would know the right action to take for it. He didn't want to be the reason for a bad day—not on his watch!

"I'll think about it," he said.

Later that day, out to grab a coffee with a buddy, Mike, from the Process Safety team, Derek talked through his dilemma.

"How can I contribute effectively to an Alarm Rationalization Study for a unit I know very little about?" Derek wanted to know.

"You know," Mike said, "a couple people from a company called Risk Alive are putting on a workshop at site tomorrow." He explained that they were going to talk about how people could leverage their platform to learn from PHAs and Risk Assessments—learn about safeguards, about recommendations, and the most critical threats in a unit.

Derek had sat in one PHA years ago, thinking that these studies were pretty awesome, packed as they were with a bunch of subject matter experts sitting in a room and brainstorming what could go wrong in a unit, asking what we had today to protect us against it, and theorizing on what we needed to do to close gaps that would make the unit safer. But he also remembered those PHA reports being massive, hundreds and sometimes thousands of pages long! When he mentioned this part to his buddy, Mike made it sound like this new Risk Alive platform could actually distill all of that information in a way that would allow him to pull out critical learnings relevant to his task. This piqued Derek's interest, because if that was actually true, maybe there was something they had that could help him with his problem of Alarm Rationalization.

Derek asked Mike to send him the invite and joined in on the workshop the next day.

At the meeting, the Risk Alive platform looked intuitive

to navigate on the surface, and they had some neat graphics and visuals to communicate risk learnings from PHAs. The presenting team got to the section in the platform that gave you a prioritized list of safeguards for a unit—all based on risk! You could even click on a safeguard, and without digging through hundreds or thousands of pages, you could see the exact conversations that the PHA team had around that safeguard: threats, impacts, risk levels, and more. This got Derek excited enough to raise his hand.

"Can I use this to get a list of critical alarms in a unit?" he asked.

The answer was a resounding yes! At the click of a button, you could filter the entire list of safeguards to just show you the most critical alarms for a given unit, and with a second click, you could learn why that alarm was important and what it protected us against. This was precisely what Derek needed to know to be confident going into the Alarm Rationalization exercise for the Poly Cat Unit. It was true that he hadn't spent much time in the unit and didn't know its alarms, but it didn't even occur to him that he could use data and the experience and knowledge from the PHA team to learn more. With a platform like this one, he wouldn't need months of time to learn how to use it. Just two clicks and he's there!

Immediately after the meeting, Derek gave the Area Operations Manager a call and told him he was in.

"You can count on me for the Poly Cat's Alarm Rationalization Study next week," he said.

"Excellent news!" the manager said before pausing as if in thought. "But what changed your mind?"

"I found my confidence."

"That was fast."

With a laugh, Derek explained how a new tool meant that he wouldn't be going into the study with just his own knowledge and experience; he now had hundreds of years of subject matter expertise, knowledge, and unbiased data to back him up.

Fast forward to the next week, and there he was in the room for the Alarm Rationalization Study. The team started talking about an alarm, and inside the first five mins, someone said, "I wonder what the PHA team said about this alarm."

"Let me go pull that PHA from the vault shelf," another team member said with a grumble.

"Hold on," Derek told him. "Just give me a minute."

They watched as he calmly took control of the screen, pulled up the new platform, and looked up the tag. "There we go!" he said, pointing at the answer right there on the screen.

There was silence in the room. Everyone looked at Derek like he was the smartest guy in the building, and he experienced a feeling he hadn't expected. From that moment, the discussion in the room evolved. They were referencing the PHA team's insights through the new platform for every alarm they talked about, and at the end of it, they could all walk away knowing that the facility was safe. If an alarm went off while they were asleep at night, anyone in the control room could look at the documentation and have the best, most data-driven intelligence to act upon. They knew that anyone in the chair at the time could feel just like

Derek did: that they could run the facility in the best and safest way possible.

Night and Day (Before and After)

The story about how that easy and open access to data changed the way Derek thought about his first career didn't end with that Alarm Rationalization Study either. The very next day, he learned how this information opened up whole new worlds of impact opportunity.

Derek now had access to any tag number and all equipment in the plant that was safety related. It felt like he was walking around with a new flashlight, one of those ones the police use, the ones with the narrow beams that shine so bright it's like you can see the future. With his new and powerful flashlight, he began snooping here and there, being curious just because he could. A squeal of a bearing on a pump: *What noise was that?* he wondered. A strange, sweet smell wafting from a cooler: *That can't be good.* In a new and profound way, Derek felt empowered to *do something*.

Crouching down, he found the pump tag number and looked it up in the database. The search was returned with a series of potential threats to mechanical failure based on high temperature and a seal failure. He checked out the pump temperature with an infrared gun, a gift from the maintenance guys a few months prior. Sure enough, the temp was high—higher, in fact, than the product temp in the pipe. In response, Derek quickly started up the backup pump, then emailed the safety-related text about all pump failures listed in the safety-related data he'd been given

THE SOUL OF RISK

access to, linking the message to any pump with that same tag number.

Satisfied that they'd averted disaster, he moved on to the cooler with the weird smell. The sweetness of the odor had him suspecting a glycol leak, but he couldn't be sure. So he entered the tag number into the system and inquired about leaks. Sure enough, on the exchanger side was "glycol," and the threat of "corrosion, tube leak." The cooling louvers had also failed, closed, and were operating with high temperature. One of the impacts was "over temperature on glycol loss." Derek copied the text in an email to maintenance about what he'd found and diagnosed. Another crisis averted.

His spider sense grew as he wandered around the site asking himself, "Anything new here? Anything unfamiliar? Out of place? Missing?" With each suspicion, he knew he could diagnose a potential problem using this new tag-based access to all safety-related information.

Derek had time during his inspections and rounds to be proactive, and perhaps prevent the next surprise or near miss. He understood the impact of unsafe days and reminded himself when the site had burned up one of their boiler units and at the same time managed to burn up a cable tray carrying signals from many areas of the site. That one surprise and unsafe day resulted in a six-month outage to rewire and rebuild the boiler units and instrument cabling system. It was Derek's boiler unit. He was questioned and felt responsible for bypassing some of the boiler controls that had a history of failing—and on his shift and under his authority advising others to do as he had done before. Since that surprise of a day, seeing risk clearly had been paramount.

So that day when he first began wandering around the site with full access to all this knowledge was a special one. It validated that Derek was able to contribute to something bigger. His purpose in this organization had leapfrogged his own expectations. His perspective on this organization had changed forever. Now his view of what was possible for himself and those around him had evolved into something new and exciting. Derek was becoming someone who mattered more than he thought possible. He had started as a young engineer looking to learn and discover how to matter and bring value to a team, but going forward, he had a new sense of clarity on how to keep his site safe and reduce its spending.

Your Help is Needed

A career in Risk & Safety is a career in learning. Just wanting to be "safer" is such an immature thing for an organization to desire in this day and age. It's so far behind the times to think that only "safer" is possible, instead of simply *safe*. Being *safer* isn't good enough when being *safe* is so important for industries, the environment, the corporations owning the hazardous sites, our colleagues and friends who work there, our kids who work there, you who work there.

Even with all the data available to us at the touch of a button or the tap of a finger, the perspective on *safe* is so unexplored and not thought about with enough clarity and perspective. It has been so ignored and so mistreated that it usually appears as little more than an expense line for executives, managers, and corporations to approve every year. Although the words Risk & Safety often appear in vision and mission statements to portray a core value (which they may be), Risk & Safety are not treated as a science, or something with R&D attached or something with ROI calculated monthly. Risk & Safety, for the most part, has been dealt with as behavior, compliance, regulatory, and a legal matter by every business.

But we can't keep ignoring this shortfall. Every day, week, month, year, incidents and near misses remind us that we are *not* in control. Sometimes they make it clear that we are more like *out of control*. The matter of keeping us *safer* has been appointed to a single person of authority with essential staff who are by themselves supposed to ensure that compliance and regulations are met, along with incident and near-miss investigations into accountability. But let's not be stupid here; there are better, more proactive ways to go from *safer* to *safe*.

At most organizations, Risk & Safety is so scary and hands-off—for the most part, unchanged for forty years at the foundation level—that the approach to being safer has also remained unchanged (with the only exception coming through loss and incidents shared so kindly by insurers looking to grow their bottom lines).

That's the bad news on an industry level. On a prospective first-career level, this news could not be better, as it leaves the runway for learning about Risk & Safety wide open. First career people are poised to enter the workforce with the freshest perspective, a new understanding about how best to tool up with AI and ML. The young people entering the industry as a first career are in the best position to help this siloed, tattered, practically self-doomed field.

The chemical industry alone is fractionated beyond 10,000 risk consultants working in the 250,000 chemical handling and manufacturing facilities worldwide. Then there are the other industries, like pulp paper, nuclear power, pharmaceuticals, agriculture, and on and on. Yep, in all industries, we are positively *pregnant* with the thinking, "We are smart, committed, and experienced, and if we go with what we've always done, we're good."

For all the young people out there considering Risk & Safety as a first career, know that experts in this field need to be convinced that it's time to drop that sense of pride about a great safety record meaning nothing more than just hitting the reset button and feeling "pretty good" about avoiding unsafe days. Those experts need to understand that it's not about the safe days you've had, but about the risk going forward. And to help with this new understanding about risk, we need your tools and the science you grew up with.

You can't *see* risk. Risk can be assessed, evaluated, understood, and controlled. It is expressed as the probability that a person, the environment, or assets may be harmed or damaged and would suffer adverse negative effects when exposed to a situation that can cause harm or damage. But now, the Risk & Safety business is like the wild west, inviting the curious and brave to be pioneers and futurists. If you're excited about a career enabled with science and data, then join an industry where you get to apply both, thick and heavy, to the information compiled by each facility worldwide, those facilities just waiting for someone to give them a reason to work together.

With your help spreading the word, the available data can make us all safer—not just every site, but every industry, and in fact, the world as a whole. The runway for learning is wide open. Risk Alive recognized it seven years ago, and now we seek the brightest minds to help us deliver it.

I like to use the acronym RAIN, first coined in *Awakening Joy: 10 Steps to Happiness*, by James Baraz and Shoshana Alexandra. In this context, I have changed the purpose of the original acronym to suit the context of the message at hand, but the beauty of RAIN is so powerful that it fits well here.

Put simply, RAIN is a change of state. It wakes us up, gets our intention, helps us be grateful, and be alive. It represents and enables our planet to grow, flourish, and serve others as they are born, much like learning helps our industry and our world. It's so fitting for the subject of first careers in Risk & Safety. RAIN, the word, could have a long-term impact as sustainable and purposeful as Apple in its pioneering days of the first computers with 48K of RAM. For all of you first-careers, RAIN can be your approach, as you are young, unblemished, and among those eager to learn and grow, mature, and flourish.

RAIN is:

(R) Recognize	Recognize the strengths you bring	Seek purpose in what you bring
(A) Allow	Allow life to be present	Be present in every moment, be curious
(I) Investigate	Investigate Ideas thoroughly	Seek clarity, and a view from the top
(N) Negate	Negate the opinions of others	Know it's not about you. It's about them.

Take a few moments and try this approach. Use RAIN to remind yourself how powerful and impactful you can be, and how much joy you have yet to discover. As you prepare to enter your first career, you are truly blessed.

What's the One Thing?

We know that Risk & Safety can impact all of us, and whatever can be done, and whatever is discovered or invented to improve it, is good for all. In certain countries, it's the law. Yet why is it so untouchable in certain industries? And in some countries, and why does it take so long, with compliance and regulations lagging years behind the technology and data available?

Is there something that each of us can control as we attempt to leapfrog these viewpoints? What's the one question we need to ask?

And I'm not posing this rhetorically. Recently, I was asked this very question by Claire, a young first-career who had joined Andrea's team and had asked a similar question of a room full of experts.

The silence in the room made us (the experts, managers, execs, engineers) feel uneasy and guilty that we hadn't done enough to be safe. This young person hammered us with questions on Risk & Safety. *We haven't seen an unsafe day or near-miss for at least two years,* we were thinking. *Who does she think she is anyway?* Was it some kind of fearless understanding of *technology* that Claire was operating with, or maybe an *unrelenting passion* to be safer? And were we, the older ones, missing the boat?

Anyway, where were these questions coming from? We hadn't replaced senior people who had moved on. We weren't aligned on digital transformation and where to apply it. Now it seemed we were being told to move forward, change, leverage technology, or what?

Claire's tone and words were respectful, of course. She held the view that it was her place to question our ability to make a significant change in Risk & Safety, and that we weren't asking the right questions. She was questioning the older generation about failing to act, and perhaps that came from her having grown up never really knowing a time without Google and social media, a time when she was equipped to have the answer for everything at her fingertips. We had our core values of never giving up, believing in our leaders, and being all the better for it—just work hard and things will turn out. Yet Claire seemed more confident than our younger selves, more equipped to make decisions with just-in-time confidence, and less shackled by fear about needing to know absolutely everything.

Risk & Safety was not scheduled to be reviewed until the next quarter. But with Claire asking tough questions—not as a traditional expert, but as someone astute and pertinent, more curious, innovative, and willing to change and try new things—she opened our eyes to how the world as she knew it had certain social expectations, and that the environment and Risk & Safety were a big part of it. Anything less than *safe* was unacceptable.

"Maybe we need to reprioritize Risk & Safety with technology," was the prevailing sentiment.

Claire suggested that reinvention of how one looks at Risk & Safety was already happening, and that predicting the future (e.g. the Risk & Safety of our people, environment, and assets) with technology and learning about all things in Risk & Safety could be measured and predicted with Artificial Intelligence/Machine Learning (AI/ML) and data analytics.

She also suggested that it's time to be more efficient and use the periods between unsafe days and our last incident like a gift. Not just to be grateful for them, but to learn more from them and seek intelligence from our past work (PHAs, HAZOPs, QRAs, Audits, Asset Integrity Reports, etc.). She reminded us that being surprised by the next unsafe day at our site or our industry, and promising that it won't happen again was not acceptable, and in fact, it was an untruth. Something *will* happen again—not the same thing, perhaps, but if we don't change how we approach learning, we will respond the same way. So use the space and time between unsafe days more aggressively to find new intelligence to be safer today than yesterday—and get better at predicting and seeing the risk from the highest mountaintops—together.

The meeting left me wondering, was this our new reality, leaning on the young for their intelligence and mindset?

Questions Claire asked (and that we should all be asking):

- Did we consider everything that could go wrong? Did we miss anything?
- Can we continue with inconsistencies between our sites and the sites of our competitors? The most consistent severity likelihood risk reduction optimum safeguarding—is there a sense in considering what others see?

- Can we say we are happy with our profits, efficiencies, culture, and record on Risk & Safety?
- Are we spending on the right things, the most optimum safeguards, and the best risk reduction measures? The highest risk-reducing safeguards? Or are we spending on the wrong things, the zero risk-reducing safeguards, and recommendations/actions?
- Is it time for our company to realign itself with today's modern resources, the perspective and thinking of the young, the aggregated intelligence of others (data), so we can modernize and leapfrog our work processes and tools to enable future generations?
- Can we prepare ourselves for the next decade by employing the young to help us be safer, more profitable, and more efficient with just-in-time Risk & Safety intelligence and decision making?

Why wait for another significant event to change our industry? Why wait for the next Exxon Valdez, BP Texas City, slow ammonia leak at a local hockey rink, or stuck open relief valve leaking flammable liquids into unsuspecting lunchrooms?

If we care about the future of Risk & Safety, then significant change is required. And we are blessed to have the ingredients to make the change. We can change everyone from CEOs to operations so they understand that our work is perfect information (made perfect with technology) to be treated as data, with proven toolsets, and with the help of data experts to make better decisions, save money, and become more efficient without capital spending.

The fact that this book was written and is now being read is significant. That those users have agreed to share all the learnings from their data to help and assist others to be safer is also significant. It demonstrates what Claire so aptly told us: that this is already happening. It's real and present. We need to be more aggressive to reduce the likelihood of future unsafe days and protect our sites, our people, and the environment, our assets, and our industries.

What is taking place today is not some futuristic event. It's real, based on truths and the present. Others have crossed the chasm and accepted that the data leads to intelligence, in an affordable way to improve efficiency, increase profitability, and be safer. We can hold ourselves accountable and get ahead of this. It's merely a simple choice: Do we want to help ourselves and be influencers and contributors to build safer industries together, or not? It's a choice to be part of a new world, one with a new perspective of Risk & Safety made possible by our young people and their relationship with technology and data. Just being part of this new world is exciting.

CHAPTER 5

How to Transform the Risk & Safety Role

Recently, I had an opportunity to check in with a former operator at our organization, Joe, who had spent the past two years pursuing a second career within the company as a Risk & Safety specialist. He didn't know this yet, but I was about to promote him to a role overseeing all of our sites. First, though, I wanted to get his impressions on how he felt about this second career of his.

"Do you remember that discussion you had with Andrea two years ago?" I asked him.

"You mean when I asked her about how I could help with the Risk & Safety initiative?"

I nodded. "Back then, I'd been thinking about you as someone who might be of great help to the team. You've always had that ability to inspire our people, especially operators, to speak their minds. I just wasn't sure if you'd be interested in following the data to help us create change."

THE SOUL OF RISK

Joe grinned and scratched the back of his head. "Yeah, I have to admit that that part seemed pretty foreign initially. But then, the more I worked with it, the more it gave me a sense of greater purpose."

"So this second career has been a dream come true?" I asked with a smile.

"Honestly, yes. I enjoy working with others, providing them with the data they need to make decisions with confidence. It's been really freeing to have data that helps me avoid supporting ideas designed to just keep production running. And I love it when I see operators standing up for what's on their mind and being curious about what the data says."

Joe went quiet for a moment as if deep in thought. I nearly cut in to tell him about his promotion, but then I held my breath because it seemed like he had something he really wanted to say.

"I remember the first time Andrea challenged me when I was an operator and a member of her team," he said. "The discussion initially made me feel uneasy."

"What do you mean?"

Joe smiled. "Let me explain."

• • •

"Joe, what's your story?" Andrea asked.

"I'm not sure what you mean," Joe said, still a little unsure about how to read Andrea, this new hire who was running an entirely new unit at our organization.

"Well, I've been in a few meetings with you now, and something interesting occurred to me. It doesn't matter who's in the room—site manager, engineers, or the corporate safety VP— you challenge things on display, speak your mind, and seldom back down. You stand your ground."

"Not quite sure if that's a compliment," Joe said uneasily.

Andrea smiled reassuringly. "Yesterday, you were determined to test the low-level trip system on the boilers, even if it required shutting down the site. Nobody in the room sided with you. They all sided with the site manager, who kept probing for backup procedures that would limp us to site turnaround this spring."

Joe shrugged. "I call it as I see it."

"That's what has me thinking. I'm the one promoting how we can use all this new data, but I don't have the same working experience you have. I don't have that in-the-field expertise I can lean on. You do."

"Well, thanks, I guess. I'm still not sure what this has to do with—"

"Where do your strength and determination come from?" Andrea cut in.

Joe's eyes went wide. "I have to tell you, Andrea . . . no one has ever asked me that. Not sure I know how to answer, but maybe I could share a story that will help explain it."

Andrea gestured for him to continue. "By all means."

"When I started as an operator twenty years ago, I lost a friend—my mentor—during my first year with the company. He wasn't alone either. The company lost two others the same day."

Her face running pale, Andrea shook her head sorrowfully. "How did it happen?"

"They were exposing a buried flow line, rich with H2S

content, trying to locate the plug. The poor guy digging in the sixteen-foot-deep hole went down first. Then his supervisor, plus twenty years of experience, saw him lying there and attempted a rescue, but he went down too. Sometime after, they were discovered by my mentor, and he went down as well. I didn't find out till the next day."

"Joe, that's awful."

"It was. And what made it worse was that we didn't really even have the means to learn from it. From time to time, I would search for information about the accident online, but there was never any trace. Like it didn't happen at all. The investigation results weren't shared until years after settling the legal and insurance suits."

Joe took a deep, shuddering breath as he pondered the loss of his friend. "Several years later, I was on a night shift as an operator on another site. We had a cable tray damaged, hit with high load carrying equipment. The damage took out many alarms and trips around the butane spheres. We should've shut down, but instead, we were instructed to take on abnormal operating rounds and procedures to manually check the levels in the spheres. Someone else, the engineers and the safety experts, were recommending these procedures. And it was all approved by the VP of Safety, with buy-in from the VP of Operations.

"There was a large construction project with dozens of workers close to the spheres at the time. Just as I was leaving one of my night shifts, the PSVs on the sphere let go, spraying liquid butane all over me and the incoming morning shift of construction workers. Most of us went to the hospital." Joe shuddered. "We were so lucky there were no vehicles in the area. No static charges let go. Nothing happened to ignite the explosive mixture. It could've been disastrous."

"You're lucky to be alive."

"Exactly." Joe shook his head bitterly. "I felt like I was being reprimanded as the investigation team interviewed us several times, searching for anything that looked like failure in the backup procedures that might have led to overpressure. They were searching for answers, hinting that no PSVs would have been activated if there was no over-pressure situation. Funny, isn't it? The PSV's venting to the atmosphere is a primary safeguard, but in this case, it was the source of the hazard. Funny that someone didn't think about that—that someone didn't recommend that we shut down operations and repair alarms and trips. Instead, we implemented people-based procedures to replace the alarms and trips. This went against the intended design. And it's exactly why we all had that near-death situation."

Joe took a moment to let that sink in. "I guess that's also when I started standing up to anyone and everyone about safety," he said to Andrea. "That event made me learn to care. It drove me to keep my eyes open, speak up, and not back down, no matter who's in the room."

"That's an impressive outcome."

Joe shrugged. "Safety, being a team player, and helping management get their way—as an operator, that's not your job. You're accountable to stand by what you believe in. It comes down to who you are, how you were raised, and the culture you want to surround yourself with. So, because of what happened, I always recommend taking the plant offline. Take the time to test the low trip and low-pressure trip on the boilers, ensure the integrity of the SIS systems, the primary safeguards. Question, anything recently installed, were they following intended design and recommendations by the safety experts?"

"I agree completely," Andrea said. "The two most vital things in these situations are standing up for what you believe in and following the data. If you have those things, then you can work with the experts, engineers, and site managers in a way that's based on intelligence and information rather than just crossing your fingers and hoping that you don't have an unsafe day."

The two of them pondered this wisdom for a moment before Joe spoke again.

"So that's my story, but what about you? I mean, you're practically worshipped by the executives and management at the site. How did you become so passionate about following the data?"

"Thanks, Joe," Andrea said with a smile. "I guess I owe you now that I can fully appreciate where you're coming from. I once worked for an owner-operator. We'd gotten in trouble with a regulator who pulled our flare license, citing that we didn't know our risks. In response, we sought some help from a consulting firm with a background in risk data and analytics to help us. We needed a quick turnaround to help convince the regulator that we understood our risks and were managing those risks. Without that evidence, we wouldn't get our license back and be able to start back up.

"We found someone with the credibility to help us, and they quickly requested our data on our PHAs, testing and maintenance programs, permitting systems, and our

Business Planning and Control System. Within a short time, they ran the analytics on this data to determine what matters most, the most critical threats and safeguards. Then they conducted site interviews." Andrea was scrolling through her phone as she spoke, and then she paused to show Joe her screen. "Here, look, I found this snapshot. It's a result of what they came up with."

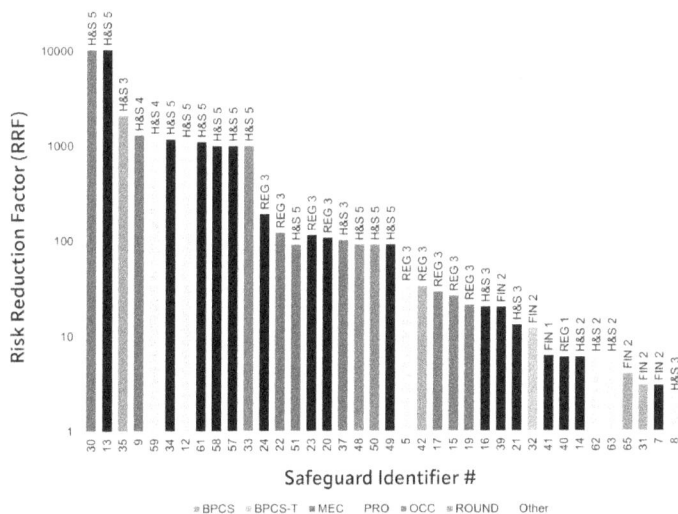

Ranking	Safeguard Description	Safeguard Tag#	Invalid Safeguard Justification	RRf
1	MEC - PSVs on Hot Carbonate Turbine discharge set at 123 psig will relieve to atmosphere., MECH	13	PSV-123/456 had simultaneous work orders for maintenance.	10182
2	Other - Sulphur plant burst disc/bypass line will protect Amine Still., OTHER	34	Burst disc flange leak.	1170
3	Other - Hot Carbonate Pump Procedure to open valves in correct order., OTHER	61	PRO-123, step 6 (pump startup). Wording is "to prevent lifting the 50# relief valve". NOTE: Shutdown procedure doesn't mention an order. (PRO-678)	1091
4	BPCS - TAL-2500A to H alarm with operator action., BPCS	58	2500 A through H are set at max range to prevent activation.	990
5	Personnel in Amine Still Condenser area less than 10% of time	57	Operator action is event of process upset is to investigate in the Amine Still Condenser area. Therefore, this modifier is not valid.	990

Joe nodded in approval.

"The company quickly proceeded to educate, fix, and demonstrate that the most critical safeguards were in fact now understood, fixed, and working," Andrea explained. "And then we shared the changed Risk Management programs with the regulator, showing exactly how we would keep track of these critical threats and safeguards going forward. The consulting company we were working with suggested they could share information with the regulator on demand, always using the most recent Risk Assessments to validate the criticality of threats and safeguards. Evidence of the Management of Change would therefore always be current, in service, and accurate."

"Huh," Joe said. "So it's like you had your proof of concept."

"Exactly. And anyway, I have personal reasons for believing in the power of following the data."

"Personal reasons?"

"I have MS. Wasn't supposed to live past my early twenties. But now I'm into my forties, and it's thanks to those who share their data, data science, and technology to find new treatments for my life-limiting condition. If it wasn't for those research people following the data, I wouldn't be here and wouldn't be contributing to this company today."

• • •

The day after this conversation with Andrea was the day that Joe first approached me and expressed an interest in contributing to our organization's new focus on using available data and technology to improve our Risk & Safety perspective. His site's budget had already been increased thanks to Andrea's perspective, technology investment was improving, and Joe started thinking it was time to do something different.

"It just seems to me that talking about change is one thing," he explained, "but actually taking action on this plan will require quite a bit more from us."

Joe, of course, was right. People in any industry have a tendency to get satisfied—or even complacent—with the way things are; at least until they have an unsafe day, or their neighbor down the road has one of their own.

"What did you have in mind?" I asked Joe.

He explained that his idea was for him to assume a second role within the company, continuing as a site-level operations manager primarily but also working on and adding his unique perspective to this new organizational focus on Risk & Safety.

"So you're telling me you want to do two jobs?" I asked wryly.

Joe smiled. "I'm thinking about this as a new opportunity. It'll broaden my skills, and my value to the company."

This was music to my ears for more than one reason. First, it was inspiring to hear that someone as skilled as Joe wanted to branch out and contribute to this new initiative. But just as importantly, I had been having suspicions lately that Joe had been planning to leave the company. Like every other company during that era, we had been experiencing extremely high turnover that year, and losing Joe would have been a major blow. Now, thanks to Andrea's innovative thinking, one of our best operators seemed to be invigorated by the idea of getting out of his usual box and helping the company with new things. Of course I told him he could start immediately.

"What do you think I should do first?" he asked. "I mean, I have some ideas. I'm just not sure whether it's my place to implement them without approval."

"The first place to go is to Andrea's office," I said. "She'll have some thoughts on what direction you can take this. But I can tell you right now that she's going to be thrilled to have someone with your skills and site experience taking the lead on this. We've been talking about recruiting new people to fill these roles, but who better than our own people? Especially when it's someone like you."

This seemed to put a bounce in Joe's step as he headed down the hall toward Andrea's office. Later, I heard from Andrea that Joe approached to ask her what she thought. He didn't want to be presumptuous, because after all, it wasn't necessarily his job to offer views on how to be safe.

"We have experts and engineers who designed our facility and are implementing the new technology," Joe told Andrea. "So maybe you already have all the input you need."

Andrea said her eyes went wide as the full weight of this idea sank in. Like me, she had been considering this Risk & Safety initiative through the lens of hiring new people from outside the organization. But here was Joe, a man who faced risk on site every day, a man she had just spoken to about what a standup guy he was. Who better to help launch Andrea's newly approved project, leverage the technology and data, and make us more efficient, reliable, and safer on a site level?

The question sparked a memory that Andrea shared with me later. One of the slides in her initial presentation had depicted a site demolished by a massive explosion, and in the foreground stood an operations manager sorrowfully observing the fallout. She'd put it right there in her presentation, this subliminal answer to our questions about who in the world we were going to get to spearhead these Risk & Safety initiatives on a site level. Joe could not have been in better position to not only help us learn from the unsafe days he had experienced firsthand, but to help us get ahead of the curve and use the data to avoid future unsafe days. In Joe's hands, the data would be more than just numbers; it would directly inform the significant risks happening on site, and show us how to address the threats, impacts, and safeguards both in place and desperately needed. And who better to stand up for the data than a guy who had clearly demonstrated he wasn't afraid to tell it like it was to people at all levels, both inside and outside the organization?

"This is brilliant, Joe," Andrea said.

Joe was over the moon about the opportunity, but he also had some reservations. "I may need your help to understand what some of the data findings mean," he admitted. "I'm new to this kind of thing."

"We're all new to it," Andrea assured. "I'm happy to help. And if there's anything I don't know, I'll pass it on to the experts to investigate and find out what we need."

Tasked with the authority to lead the Risk & Safety initiative at his site, Joe got straight to work. He soon picked up on some important, unforeseen concepts related to his facetime with risk, discovered ways to apply the data to fix the problem, and was learning like crazy and having fun in the process. What began as a simple question about whether he could help with Risk & Safety quickly blossomed into a new career focus for Joe. Each month, he spent more and more of his time contributing to Andrea's team. Eventually, two years into the role, he got to where he was so good at this work that the executive team decided it was time to promote him, and I was the one who got to deliver the happy news.

"Joe," I told him with a smile, "we're moving you up to the bigtime. You've clearly shown how valuable you are on the site level. But I'd like to offer you a promotion to Safety Manager."

"What does that mean?" Joe asked excitedly.

"It means we'd like to move you from site to site to help address each plant's unique needs."

He looked down at his shoes humbly.

"You in?" I asked.

Joe broke into a wide smile.

From the perspective of a man who cares about safety, I no longer worry about whether we'll have any unexpected accidents at our sites, because Joe has all the qualities we need to ensure that we're always following the data and standing up for what is right and necessary. And from an employer's perspective, I no longer worry about whether we're going to lose Joe like we've lost so many other talented employees over the past year. In fact, it's been a long time since I've seen an employee so satisfied with his job.

The Operator and Facetime

As it turned out, Andrea and I were right: Joe could not have been better suited for his second career in Risk & Safety. This is often the case for operators, in fact. The question is why. What makes these people so adept at the job?

Well, first off, they're naturally in a position where an adjustment in perspective can lead to the most profound change for the organization as a whole. For most operators at just about any site, there's a tendency to favor profitability over safety, at least to some degree. And it's almost an automatic response. Back when Joe was an operator, whenever equipment failed on any given day, he would put the bypass on quickly—and he wasn't alone in this; he had witnessed and worked alongside others who had done the same. So had I, back when I was an operator. The instinct is to keep the sales meter running, to cross your fingers and just hope it works to keep the production systems up. In my case, doing what Joe and every other operator tends to do never made me feel *bad* necessarily; just maybe a little afraid about what could happen with the inlet valve on bypass. Mostly, I would just hope that the bypass would survive long enough that there wouldn't be any more hassles on my shift.

You can see where this is going. One day, my luck ran out. A slight chattering of a pressure relief device on a hydrocarbon system meant that a key safeguard had failed. If we wanted to fix the PSV, we would need to shut down the whole facility. But we didn't want that. The collected opinions and experience of most of the operators on site that day said the same thing: it would be safe enough if we just isolated the PSV and sent it out for repair.

"We've never experienced high pressure on this circuit," was the reasoning.

The arguing that followed was, to put it mildly, intense. I was one of the operators in the camp that did not like this idea. This plan did not seem at all safe enough to me. *Then again,* I thought, *who am I to question my peers anyway? That's not my job.* The moment this thought passed through my head, I pictured my grandma scolding me for blindly following along with what others told me to do.

I'm sorry, Grandma. I wound up doing what others told me to do.

But here Joe had seized a new opportunity to avoid future unsafe days in a way that my site failed to do. Today, if Joe finds himself facing a situation similar to the one described above, he has access to the data about why that specific PSV exists and all the hazard events it protects against. As a result, he doesn't have to simply agree to isolate the valve and continue operating. He's able to present his case in a way that sounds more intelligent, doesn't come across as just a gut feeling, and is fully supported by verifiable, data-driven accountability. Armed with this data, he has far less trouble convincing his fellow operators to do the right thing—all of this made possible by learning that the data from our past PHAs (Process Hazards Analysis) can be repurposed.

In this way, Joe proved to be an even better investment than we had hoped. He advocated hard for building what we called a "Hunger to Learn" team, a group of operators looking to gain a new perspective at his site by using an AIP (Aggregated Intelligence Platform) built on thousands of PHA data sets. Thanks to the work of this team, our organization learned that when equipment fails, the data provides us with a new ability to make the best decisions, and the lowest risk decisions, even as we *keep operating*.

For Joe, it became a new edge in his professional life. It spawned a new career in Risk & Safety that was supercharged by the data.

"You know, it's funny," Joe told me recently. "In my previous job as an operator, I always felt that the right thing to do was make decisions in consensus with those who came before me."

"Like me?" I asked.

He nodded.

"Why in the world would you think it's a good idea to repeat the decisions *I* made?"

"Good question."

We shared a laugh. And then Joe went on to explain that his quickest and most profound lesson learned was that data matters because it always helps us make better business decisions, even in the Risk & Safety sphere.

"Having access to the data and the AIP grew my confidence," he said. "Every day at work between unsafe days, I followed my new obsession to figure out *why* we were safe that day instead of just *hoping* we would stay safe until the end of my shift. I would get to that *why* by analyzing our top threats and comparing them to impacts at similar facilities. Through this work, I was surprised to discover threats and impacts I hadn't considered, along with extra safeguards that we weren't using. I became fascinated with other facilities and their data, and I wondered about their experiences and safeguarding strategies. At my site, we hadn't thought about numerous significant threats and impacts before. Being proactive rather than surprised by an unsafe day changed everything about how we worked."

The Power, Purpose, and Strength of PHA Data

Joe's journey from operator to Safety Manager is not one that has to stand alone. Any operator at any site in any industry is capable of replicating what Joe did for us. And the best part is that it's a win-win. The Joes of the world get to demonstrate their ability to quickly learn, improve their attitude/self-worth/pride, and build into a second career in a new field. And the employer wins because the Joes of the world are so perfectly equipped to use the data to improve operations and reduce unsafe days.

I know all this because I led a similar path in my own career. Back when I was working as a site engineer—just like the 250,000 engineers at 250,000 sites worldwide—I held an unchallenged perspective about PHAs. Often, I was asked to organize and coordinate the PHAs because I had access to the hazard information they required. That was really it. No other expertise necessary. I began the process by pouring through the reports, which had been collecting dust on the shelves since our plant started up some thirty years prior. Hidden in those reports, I found hundreds of clues and learnings about being safer. But the most important thing I discovered was that these learnings could give anyone (as long as they're willing to look closely) a new level of power, purpose, and strength.

Of course it helped that I have a natural interest in these kinds of things. I quickly found myself taken in by the prediction models, whether using raw PHA data or refined learnings. The health checks that I could generate with the prediction models were always particularly enamoring to me. The PHA data was (and is) just so full of learnings. And in an industry that has for decades relied on gut feelings and luck, it is so refreshingly *predictable*.

We used the resultant AIP to prioritize our safeguards, find our inconsistencies to improve PHA processes, understand the best sequence of recommendations, and see the risk profile changes according to the sequence of implementation. It also led to a pride-check, which in my view, was the best part. We had to swallow our pride and discover what threats and impacts we missed and were not considered in the risk assessment of our site. In retrospect, it wasn't that difficult to accept what we didn't know. After all, how could one site possibly know everything that could ever go wrong? It simply could not.

Until now! With the technology and databases connecting all the world's data, every site can know everything that has gone wrong at any similar site anywhere and everywhere in the world. With this data compiled and accessible to all, any site taking advantage of it can be as safe and profitable as possible.

In *The Catcher in the Rye,* J.D. Salinger wrote, "The mark of the immature man is that he wants to die nobly for a cause, while the mark of the mature man is that he wants to live humbly for one."[6] Well, in the spirit of that inspiring line, it didn't take me long to become a full-fledged member of the "follow the data" movement to make the world safer. And I'm proud to say that it's a movement Joe and many other operators like him are joining.

I challenge anyone with a job relating to site risk assessment and PHA to think about this: What do we really know at one site, even as a dedicated group operations, maintenance, and engineers wanting to do the right thing and understand what can go wrong and the best way to mitigate the risk at all times?

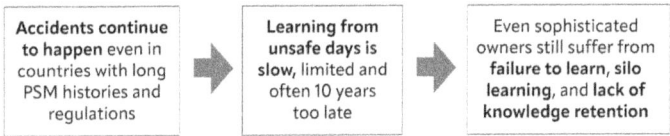

Accidents continue to happen even in countries with long PSM histories and regulations	→	Learning from unsafe days is slow, limited and often 10 years too late	→	Even sophisticated owners still suffer from failure to learn, silo learning, and lack of knowledge retention

---

6   In the novel, Salinger's Holden Caulfield ascribes this quote to Austrian psychologist Wilhelm Stekel, but the provenance of the quote has been in question. The earliest version (written in German) is attributed to the dramatist and novelist Otto Ludwig.

Have you ever felt like the picture above? Or have you seen someone wondering, "What happened? Why us? What has to change?" Maybe this picture becomes a safety moment for anyone participating in a PHAs and Risk Assessment. Perhaps it's a reminder of why we need to consider the data on a world scale. Maybe we need to stop fooling ourselves when someone says, "We got this. We know best."

The world bank of PHAs started in 2015 and continues to grow daily. Soon, it will reach 100 million learnings. This AIP provider saw the benefit of generating all PHAs into one. It's something every site can use.

Now when you ask a person to participate in an upcoming PHA, you can say, "Join us to share your experience and help us turn that experience into clues and learnings for our site and all sites worldwide. Your work and this PHA will be repurposed and played forward to the world. Do you want to help ourselves and the world to make better decisions on bypasses, work permits, maintenance, training, and get competent at keeping our risk low? It's now good for all because of the data and you."

If you're considering a second career in Risk & Safety, consider this: not only was it gratifying to repurpose our old PHAs and apply the data; it was *super interesting*. Today, I work with clients worldwide and find myself challenged with new problems to solve every day. I share my data experience, pay it forward, and am continuously learning. There are powerful, inspiring tools at my disposal, and my career and personal growth has been extraordinary as a result.

The Manager and Facetime

Not long after Joe accepted his promotion, I happened to be visiting another site up the road, and they asked me how they could replicate our success. They didn't have any operators like Joe who they felt they could easily position to lead the initiative and follow the data. I told them that, just like operators, managers are perfectly suited for a second role or career in Risk & Safety.

The leaders of this company loved this idea so much that they immediately introduced me to Sheila, a site manager they believed would be in a strong position to lead their own charge. Even in our brief meeting, I could tell that Sheila was a strong candidate. She clearly possessed the accountability, the wish to find something she was really good at, and the desire for a new challenge (and maybe the promotion that could come along with it).

When I asked her why she thought she might be interested in following the data to lead her company's Risk & Safety initiative, she started by outlining her typical day in her current role as site manager. "I work a lot with budgets," she said. "So I'm always wrestling with questions about spend and how to cut."

She went on to explain that, for years, she defaulted to what others had done before her.

"Why?" I asked.

She shrugged. "I guess I just saw it as part of my job—like reimagining the budgets was outside my skillset."

Just like with Joe, I could personally relate to the challenges Sheila was describing. In my early days as a facility manager, I struggled mightily with budgeting, estimating, and forecasting. Actually, I *sucked* at it. And the worst part was that this had to be a regular activity for me—monthly, quarterly, and annually. One year in particular, my pride was hurting. I let my staff down while using their input to produce the budget, and then it wasn't approved, forcing us to tweak it again. Given the headache we'd all just had to suffer through, I desperately wanted to inspire my staff about the upcoming year. They worked hard and generated a lot of good ideas. It was the basis for next year's budget. What didn't I see?

A large part of our budget allocation in our newer facilities went to supporting new technologies. We learned from the data retrieved from the AIP that over fifty percent of our risk reduction used complex technology such as SIS safeguards. Hence, we carried a lot of high-tech spares and needed a lot more vendor support to help us whenever safeguards would fail. In the past, the expectations by both corporate and others always felt that we should make do with less because we had newer technology. But the cost of maintaining that risk reduction was much more with parts and support of such sophisticated systems. I remember thinking, *It's not my job to question the PSM practices, or ways of the past. I shouldn't assume that accepting traditional approaches is the cost of staying safe.* But then again, was it possible we had too many safeguards or had spent too much on the wrong things?

I related all of this to Sheila, who nodded knowingly.

"We discovered that we had a lot of zero-risk-reducing, expensive safeguards and expensive recommendations coming out of our PHAs," I said. "This made us realize that we could reallocate our budget by eliminating some of the zero-risk-reducing safeguards and recommendations."

In my case as a manager, the older plants relied more on procedures, operator rounds, and inspections, and their parts were less expensive. By mirroring their practices, we reduced our spending on these actions by fifty percent. That freed up the additional budget required for other things, like spare parts and vendor support.

The data on our top ten safeguards was especially remarkable, as it showed that none of them were the expensive SIS safeguards we had recently installed. Put simply, the exploration showed that we were underutilizing these SIS safeguards for certain hazard events, and if we could change our processes to use them more efficiently, we could eliminate some of the expense.

Additionally, the data showed that the PHA recommendations we had planned for our twenty-eight process units were not optimal for risk reduction. By optimizing the recommendations into the ultimate risk-reducing arrangement, we could eliminate the need for thirty percent of the recommendations.

"That's amazing," Sheila said. "Were the budgets approved?"

"Fully," I said. "And the leadership team showed excitement about an upcoming year where we could enjoy the reduction in spending without hurting our ability to be safe."

Sheila nodded, and suddenly, she looked more and more in line with the idea of following a similar path toward a second career in Risk & Safety. This was good news, because she was clearly the perfect candidate from the company's management pool for the role. I could see that she would serve them well.

Confirmation of that suspicion came soon after. Sheila implemented a similar "follow he AIP data" method at her site, and soon, other plants soon got wind of what she had done and copied the approach. The method became the corporate standard. Across the board, plants were using the data to understand the criticality and actions to make budget allocations more streamlined and adjust those budgets on a risk-discovery basis. It seemed that a movement was born.

Data Intelligence is a Journey

If your career has been anything like Sheila's, you might have heard something like, "We manage our maintenance budgets in such a way as to reduce surprises across our business units and keep our most important assets reliable and available, ensuring maximum production and uptime." But have you ever heard this one? "By identifying which safeguards are near zero-risk-reducing, and then replacing them with more profitable safeguarding strategies, and ensuring the use of fewer human-dependent safeguards, we enable higher and more predictable profits and uptime?"

Or how about this one? "We conduct PHAs every year on our facilities, which is obviously more frequent than the minimum five-year OSHA standard to review PHAs. Our PHA teams are knowledgeable, they seldom waste our money, and they tend to find something new and correct it every time." Maybe instead, it would be better if it sounded like, "We analyze our risk using OpenPHA, seeing clearly the impact of bypassing and running without safeguards. We see no need to wait five years to redo a PHA, ever."

Okay, so then consider the organization that has a team of experts that examine all risks and dictate which hazard scenarios need to be reviewed on which process units. Now imagine that same organization hosting a digital library that tracks all their risks and PHA results with open access to everyone in the company, enabling everyone from the operator to the CEO to clearly see all risks. Instead of dedicating so many assets to a Risk Management program that identifies and deals with only the organization's highest risks, what about using predictability models on a worldwide PHA database and running health checks on all assets at least once per year? Would that not help the organization to better understand their highest risks and where they can improve efficiency and streamline spending?

For as long as business has been business, management has struggled to find the budget to implement new recommendations or justify them. With AIP data intelligence at hand, there is no need for escalating budgets. In fact, aggregating or replacing safeguards with inherent, technical, or mechanical safeguards are now within any organization's reach. Now management can enable a continuous reduction of safeguarding costs. The longer a facility has been in place and following PSM standards, the greater the opportunity to save money.

Think about it: How many people-based safeguards (e.g. alarms and operator rounds), less dependent than inherent safeguards, remain unknown and unreliable until the moment of truth, when an operator finds himself face-to-face with a threat early one Sunday morning? If this is the reality—for any organization—are you really as safe as you can be? Why take such an enormous risk? Why not do something that will save money today and increase the likelihood of one hundred percent risk mitigation by replacing an unknown with existing and aggregated safeguard strategies?

The journey to a better future that utilizes data intelligence will rely on three keys. First, there is the resistance to real change by executives, subject experts, and organizations as a whole. We in the Risk & Safety industry must work together to overcome this resistance. The second key is that we must recruit new people to the cause, seeking out those who are curious about the data and driven to help others be safer, make better decisions in operating scenarios (whether it's unsafe days, compliance days, training days, maintenance days, or spare parts days), and improve their spend decisions. The third key is that we continue to explore and discover intelligence through the data. For all of us, the journey will be rewarding in itself, as it will lead to new perspectives that will enable new thinking.

Refilling the Talent Pool

During the rewriting of this book, this section was almost removed. But then my team and I saw things differently. We saw this section as highlighting key opportunities for new careers and businesses. Opportunities to change what we might have done before, and to rethink how we might solve these kinds of problems today.

What if we deployed technology and new work processes in a way that allowed businesses to function with fewer experts—experts dispersed into the field with powerful decision making tools, AI, and semantic search tools, experts backed with data and the many lessons collected from the rest of the world? Now this section makes sense, and we have decided to leave it in, because after all, it highlights why a career in this industry matters, and why startups matter.

Enough disclaimer! On to the story:

The fall of 2021 delivered us many examples of why we must be careful not to dilute the value of experience, but to me, one stands out more clearly than the rest. During the shooting of the movie *Rust,* an inexperienced armorer handed an actor a loaded gun. The accident caused one fatality and one life-threatening injury to the crew members. How could this have happened? For decades, trained professionals handled guns and ammunition, supported by years of experience and dozens of checks and balances before any gun reached an actor's hands. But as streaming services have ramped up the demand for content, the pool of available and trained armorers has become diluted. Since other staff could move up to the armorers' position a lot sooner, bypassing the essential mentorship and due diligence in the role, tragic accidents were bound to happen. Rather than leading to a call for more training and mentorship—and more qualified people to fill the necessary roles—the event triggered a call for a ban on guns on movie sets and critical questions about balancing precautions and efficiency.

The lesson we in the Risk & Safety industry can take from this story is that, as we move forward, the most impactful and essential consideration when pursuing (or appointing people into) Risk & Safety positions is ensuring that the quality of learning and information transfer does not suffer. In the Risk & Safety business, there is an opportunity to guide companies to restructure themselves as they fill their vacant roles and do so in a way that streamlines and reduces costs across all processes. While demand for new workers rises, companies may opt to promote from within and fast-track employees like Joe and Sheila to safety-sensitive roles more quickly than before. If we're going to position these people so they can anticipate potential hazards or errors in safety-sensitive areas effectively, then we must place significant value on mentorship, support, and cross-checking. The processes should be scrutinized so that pieces of work that have been trained and polished over the years don't get overlooked, causing catastrophic consequences.

The Great Resignation

The final reason to consider pursuing or appointing people into a second career in Risk & Safety is that most organizations are desperate to recruit and retain talent, and this new field has a way of doing both.

Since at least 2008, economists and news outlets have talked about the massive exit of knowledge and talent leaving the workforce. While the messages have been loud and clear for well over a decade, the withdrawal proved to be slower than anticipated, with contributing factors such as a global recession, market crashes, and housing market fluctuations, among other factors, impacted the financial ability of that experienced workforce to leave for good.

In 2021, the long-awaited wave of voluntary turnover—a wave dubbed by Anthony Klotz as the Great Resignation—finally broke. The COVID pandemic ultimately fulfilled the prophecy that has hung like an omen for years over employers' heads. In August 2021, 2.9 percent of the labor market in the US had resigned from their positions. Employees from all roles, seniority, and experience levels started to quit, consider other options, or retire. Most were in their mid-career, between the ages of thirty and forty-five years old, people with knowledge and experience gained through the time and training that their companies invested in them.

Several factors contributed to the Great Resignation of 2021. During the pandemic, many people spent their extra free time in lockdown contemplating the value of the work in their lives and rebalancing their understanding of what they wanted out of life and their careers. In this way, the international workforce developed a preference for roles that would allow for time flexibility and remote work. Meanwhile, financial support from the government allowed people to take a step back and evaluate their options before returning to the labor market.

While a high resignation rate is always a good sign of confidence in the economy, it has a way of leaving a big knowledge gap for any organization that suffers high turnover. In response to the Great Resignation, some organizations took a proactive approach and focused on cross-training and documentation as a preventative measure to maintain institutional and historical knowledge invaluable to the organization.

Wherever there was turnover across all levels of the organization, slowing the leave was significantly less effective, as the employees who stayed often did not want to maintain the status quo in their work.

At the same time, two years of pandemic-propelled technological adaptation across all levels of organizations made companies wonder, "What else does technology make possible for our organization?" In the beginning, conducting virtual team meetings was strange and uncomfortable, but after two years, it turned into the new normal of communicating. The next step was attending virtual training or compliance renewal meetings, which now could be done from the comfort of your home with multiple ways of receiving content, such as slides, recorded presentations, and quiz checks. The convenience of technology was suddenly allowing us to consume more valuable information and review it precisely when needed.

Those small changes opened the door to a new level of thinking: If technology can support meetings and training, how can it support day-to-day operations? Can we have access to safety information in the same way we might attend a virtual meeting? What if accessing this information could be done automatically with minimal support instead of churning through tasks manually?

Viewed through another lens, this means two things for people considering second careers in Risk & Safety. First, embracing a new (or second) role in this field is a great way to shake off that status quo at work. And second, it is a dynamic way to embrace and utilize the new level of technological connectedness that resulted in part from the pandemic.

We've all learned more about ourselves during these troubling times, and we've all learned ways to use technology to make our jobs more efficient and the results of our work more impactful. There has never been a better time to apply these learnings to Risk & Safety.

From an employer's perspective, promoting second careers/roles in Risk & Safety helps maintain your valuable knowledge base by empowering key people to lead the new initiative. When you invest in your people in this way, it engages them further in their existing roles, allows them to wear multiple hats, provides them the opportunity to meet new challenges and advance their value within the company, and leads to results that will help the company become more efficient, safe, and profitable. The company benefits even as the employees tasked with the new Risk & Safety role find the job satisfaction they need to remain with their employer rather than joining the Great Resignation.

Further, after an era of work-from-home, jobs like these will compel more people to return to the office on a more consistent basis. Without that facetime with risk, it's more difficult to identify where sites can improve. Appointing key employees to Risk & Safety roles sparks a culture of on-site work, one that addresses the need to train other people and perform onboarding in person. It gets people back in the field and in the office—actually *at* work.

The rewards are plentiful for both sides, particularly at a time when we face such a remarkable opportunity for positive change. Joe is living proof. It has been several years since he moved on from his operating job to focus on Risk & Safety full time. Now he is paying his experience forward to other sites within our organization, just like Sheila is paying it forward to other sites in her industry, and I'm paying it forward to sites worldwide. We're all sharing the knowledge about how to leverage this kind of intelligence in the operational world, advocating for data-driven decision making for all operators and locations. Those who pursue this second career as we did will find themselves learning remarkably valuable skills, improving their sense of pride and self-worth, and leading from the front of a massively positive change for the world.

CHAPTER 6

Fly Toward Organizational Change

Suppose you believe in the data and are willing to take on the mission to cross the chasm to becoming an organization making better risk decisions. In that case, you will need to get past the expectations of others. Their fear of change, and their perception of change, is not required here. The experts, the execs, and site personnel in your organization will need to be swayed to your side. They may perceive that all sites are safe enough, and the cost of safety is acceptable as the cost of being in your market. Others may want to keep things simple and do what has been done in the past—why change? They'll believe it's not their job to make the change and are okay with doing what is requested of them, without question. And then, others may be fearful of change and believe that *no change* is a safer route. They'll be afraid of looking bad if any failings or unsafe days arise.

Winds of Change
Using Risk Data to Make Risk Decisions in Real Time

Experts - Execs - Site Persons

Without the Data:
Dependent on
compliance,
standards, PSM
practice, experts

- Aggregated
  Intelligence
  Platform (AIP)
- Predictive health
  checks
- Safeguard Ranking
- Lab PHAs

With the Data:
Owner stories to improve
budget allocation,
reduce spend, increase
efficiency & quality, play
forward learnings

Learnings
Analytics
Asset Centric

The Chasm

Fear

Expect words

Perception

- No change is safer
- Who says to change
- Change is unpredictable

- They told me...
- Not my job
- It's always this way

- We are safe enough
- Experience base for decision
  is acceptable
- The cost of safe is acceptable

Hence the winds of change started as digital transformation projects and touted AI, ML, technology, and partnerships with Microsoft. But now derivatives of these prevailing trade winds have been created, referred to as digital winds seeking to remove the pain of learning, and nip in the bud the questions of "Now what?" "How do we train up our org?" "How much pain will it be?" and so on. The digital winds are tools designed with ease, no pain, no training in mind. Health checks, LabPHAs, AIP platforms, OpenPHA—these are tools that never would have existed without the data. These tools are like gifts from the early adopters to you as they made the transition across the chasm and found the value of the data. So these tools or digital winds were created to make it easy, remove the pain, and eliminate the need for experts at every corner. Likely, any organization can now be asset centric, with analytics and learnings on every piece of equipment owned and operated.

The perspective on the data is simply this: many orgs are not ready for change. It only takes one or two key people to stop change. But it takes all the key people to buy in and get aligned before change can happen. We know people are fearful of change. They use their expectations and perspectives to block change. They'll say the timing isn't good, they're too busy right now, not ready for it yet, even if it is a good idea.

The winds of change were not easy to stir up. At the start of this journey, those of us doing the stirring were frustrated and exhausted as we implemented the learnings and analytics on every piece of equipment and asset. We were tired from the seventy-plus meetings with the last client, sharing with every corner of the org that following the data is a good idea. The turning point for us was when we decided we should be part of an org to help them integrate and adopt the learnings from the data and build "Hunger to Learn" teams at each site.

No matter who you are, you likely agree that a new idea is a change. If that change is on top of a practice that seems like a rock, like a forty-year-old practice of PSM and Risk Management systems, then that change is likely feared and questioned to no end. If you believe in the idea that risk data is worth the effort to make such a change, you will need to clearly see the steps to succeed.

My perspective has changed since we went through this process of the first few adopters: the more significant the organization, the more fun it will be to leverage the data for them and for us. Again, it only takes one person to believe in the data to make this change.

On my journey, the frustration disappeared as the discussion went from "that's really interesting, but now what?" (leaving me with a sense of "Oh my God, do we need to do everything?") to hearing, "Can you help us set up our own Hunger to Learn teams so that we can leverage our own data?" I think we get it now.

My thinking went from "How we can help this client, because they don't know what they don't know?" to "What else can we build, and what other winds of change can we generate to help this client who stands on the left side of the chasm looking to scale the perspective, expectations, and fear within the deepest part of the chasm?" We, as entrepreneurs looking to form a startup in Risk & Safety can clearly see the other side of the chasm. We know how good it will be for the client to have asset-centric analytics and learnings of every piece of equipment and process unit the org operates. It's just a matter of having the courage to help them fly toward that change.

CHAPTER 7

Contemplating a Startup

AT SOME POINT during my twenty-eight-year career in the Risk & Safety business, I crossed paths with Flynn, a young engineer working for an owner operator the next province over. Of course, I've worked with many engineers over the years, so it's important to note that the reason Flynn stuck out in my memory is because he was particularly sharp. He had only recently graduated from a mechanical engineering program, but something about the way he applied the data to revise the processes at his site struck me as uncommonly ingenious. It seemed to me, even ten years ago, that Flynn had a bright future.

So when Flynn recently pulled me aside to pick my brain on what it takes to build a startup in Risk & Safety, I met the question with very little surprise, and honestly, with a fair measure of excitement.

"If you could go back to when you first started your company," Flynn said as we stood in line at the taco truck in the parking lot outside my office, "what do you wish you would've known?"

The scent of simmering beef and tortillas on the fryer caused my mouth to water as I contemplated the question. "Well, I guess the biggest thing is how important it is to be clear about your purpose and intention. That's true of any business, but it's especially true in Risk & Safety."

He raised an eyebrow and looked at me sideways.

"Our hearts tell us to be brave at times," I explained. "We want to be the best we can be in every moment. But how often do we take the time necessary to discover truths and find clarity?"

"I think I see what you mean."

"You can't just think it. Not if you want to succeed as a startup. You have to internalize it." I sighed and looked over the menu. There were three different tacos I wanted, but the sign said that mixing and matching wasn't allowed. "It took six years as a startup before I realized how important that wisdom is. Six years of chasing ideas before I learned to un-hurry my thoughts so I could be the best of me. Add another six years to learn to live with intention, a good-for-all approach. It took me at least that long to realize that the world can't be fixed. It's impossible. But you know what can be fixed? Me."

Flynn looked a little confused as we reached the front of the line and I motioned for him to order first. I followed with my order and then we stepped aside to wait for our tacos.

"Let me simplify it," I said. "Looking back, my purpose and intention seem so obvious. I knew the world wasn't safe enough, so I wanted to make it safer. But it took twelve years for me to find those words. That was twelve years of lost intention. So if you're thinking about a startup, then find your purpose as quick as you can."

"How the hell do I do that?" Flynn quipped.

I smiled. "First, you have to remember that your job is to make a measured impact on those around you. Safety isn't a project or even an outcome; it's something you can always do better, no matter what. Even if you lived twice as long, it can always be improved."

With that, our food arrived, and Flynn and I tucked into some of the best tacos I've ever had. He had many more questions for me, and in some ways, taking the time to answer those questions was the first opportunity I'd had to slow down my own thinking in quite a while. If it took me twelve years to arrive at a clear picture of my purpose and intentions, I'd spent the sixteen years since living and working that purpose and those intentions. So it was with some surprise that I had more insight to offer than either of us were expecting. Our conversation that day at a parking lot picnic table, the early October sun carving through the crisp afternoon breeze, got me thinking about what other wisdom I might have to offer to anyone considering a startup in the Risk & Safety business. I shared what I could with Flynn that day and have done my part to help as he has taken the leap to form his own startup. But with this chapter, for the first time, I'll align all those key learnings I wish I'd known on day one.

Your Bench Strength

"Andrea, excuse me. We haven't met. My name is Flynn. I was just speaking with Ken at the taco truck, and he suggested I come talk to you."

With a bright smile, Andrea extended her hand and introduced herself.

"I'm a contractor with your company," Flynn explained. "Up to now, I've been helping and learning and watching from a distance. Started in the middle east, and through a Facebook friend of my wife's, I was introduced to Shelia, which put me in contact with Ken."

"Pleased to meet you, Flynn," Andrea said. "How can I help?"

"Sheila tossed me some work supporting the work you're doing, following the data, but more on the audit side of the business. I'm very good at audits, since I've conducted hundreds of them proving the safeguards are in place and working. But as I was just talking about with Ken, I've never seen anyone focus on the criticality and peer comparison of the safeguards, hazard events, and impacts. Making it obvious what matters and what doesn't matter regarding risk reduction and contribution—this is fascinating to me."

"I feel the same way."

"Anyway, I'm very impressed by your team's study determining the criticality. So that's why I wanted to come meet you in person. I think that what you and your team are working on can help many others. In fact, it's sparked an idea for a business I'm thinking of pursuing."

Andrea's ears perked up. "Oh?"

"I have an idea to help every hazard-producing facility in the world. I'd like to start a consulting company that offers certification programs—the kind that can provide education-training to people who work in hazard-producing facilities. But for the startup to work, I'll need a little support from you. With your help, I'd like to transform all the other hazardous facilities with processes that leverage your team's data and tools. The company I work for currently has learned much about the world. Our industry is so much more connected than most people realize. There's an opportunity here to leverage that connectivity into something much bigger—to do something that could make real and positive change in the world."

"Wow! That does sound fascinating, Flynn. How do you propose it would work?"

"Suppose someone knows something about our processing facility by operating something similar across the pond. In that case, we have no excuses not to make ourselves aware and leverage all the learnings we can."

Andrea motioned to her office door, inviting Flynn in to have a seat across from her desk. "Tell me more."

"I can't tell you how comforting it is to hear you say that. My first foray as an entrepreneur, I struggled to convince others of my value and bench strength. But now that I've been exposed to everything you're doing here, I see my bench strength: I not only have expertise with PSM audits but with Functional Safety Audits, Preventive Maintenance Audits, Change Management Audits, Turnaround Planning Audits, and on and on. And what it's taught me is that every single contractor in the world—at least the ones who care about others and want to keep them safe—can use Process Safety Management to greater benefit."

"So how would you do that?"

"My idea is to offer certification and training, measuring people's competency in specific areas, especially where I have audit experience in that process technology, like with ammonia or nitrogen plants."

"That's brilliant."

Flynn leaned back in his chair and sighed through a smile. "You realize that I'd be doing this on my own, though, right? That I'd be an entrepreneur, doing something for the first time, working just on the belief that I can make a difference."

Andrea nodded and folded her hands on her desk. "Flynn, just so you know, you're not alone. What you're describing is very similar to how I got to my position in this company. When I started, no one else looked at the data and learned from the results. We were just being compliant with standards and relying on substantial compliance companies and their opinions on what matters and what to do next."

"That's good to know."

"The other factor is to remember just how much good you'd be doing. If doing something today, like scientifically proving what matters by following the data, just think of how unique and safer the world can be if you're successful?"

"Exactly!" Flynn said, his excitement sending him back to his feet and pacing. "We've always relied on compliance companies' opinions on what matters. But if we follow the data to accurately understand what's happening and what we could do better, it could change everything. I don't think we've ever thought about it any other way. We just went with our guts. And since we now understand how much time and spending we may be wasting, we can recognize where we may be missing safeguards, events, and impacts that matter when performing audits. We can stop spending time and effort on things that don't matter."

"You have my support on this," Andrea said. "How can I help?"

Flynn stopped pacing. "You've already helped just by offering your support. I'm not alone in this. I can see a path to piloting a certification program with your group, meaning it will never go away, no matter which people come and go, and giving corporate assurance simultaneously.

"Let's start with getting your ideas on what certification looks like. I can send over our thoughts on a draft tomorrow."

"Thank you so much for your support, Andrea."

Six months later, Flynn had produced a series of certification programs leveraging the data to be safer. Andrea called him with thanks and a request for a favor.

"Name it," Flynn said.

"We have a partnering company in the Middle East that I think needs your help," Andrea explained. "I'm unsure if I'll be involved, but I know they need your help and your company's help. I can set up a call with the head of safety and the executives of the partnering company, and you can deal with them directly. You might consider doing some preliminary work, and I could get you their data to start with. Then you and your team could present the results. If they're impressed, make sure you send them a bill, and proceed as required. Ensure that you get them certified to see this as a lifecycle requirement, helping them with corporate assurance."

The gratitude was palpable, even over the phone. "You know, Andrea, when I started as only one person with an idea, I saw myself as an entrepreneur determined to make a difference. But going home late at night, in the dark, hearing my own footsteps on the cobblestone streets, I felt very alone. I wasn't sure I'd be able to see the creation of these certifications through to the end."

"Well I'm glad you found your way through," Andrea said with a smile. "How'd you do it?"

"Hearing a little doubt made me aware to seek others, believe in what I'm doing, and see my intentions as timely and good. Early on, I needed a goal, something to build along the journey to make the world see what I could see—all that risk data, PHAs, QRAs, LOPAs, Accident Data being produced but never used as data. All just for compliance. It was just waiting for someone to come along and do something with it."

"I'm grateful that you worked through all this. And thank you for believing in our mission. It's fuel in our tank to find others to make a difference with."

"My pleasure, Andrea. And I sincerely hope that more entrepreneurs follow what we're doing, because there's just so much opportunity. It's like I said back at our first meeting: the world is so connected. That connection allows us all to be supportive of good things. All we need is the people to provide that support."

Do You Have an Idea?

"Startup" is a precious word for the few of us feeling strong enough about an idea or a cause to act and do something about it. Most of us will not be as well-known as Albert Einstein or Steve Jobs, but we can be just as purposeful and proud about becoming part of something unique and worthwhile: our very own startup. And we can be our own platform of achievers of wild success in whatever we set ourselves after. This is, of course, assuming that we can ignore the negative influencers—from those we have never met, to our own friends, family, and colleagues—calling us crazy or hearing it won't work.

It's hard pursuing something in the face of negative feedback. The bad opinion platform is widely available on Twitter, Facebook, and Google. It helps the world judge itself by the hour, and no one can escape. Becoming famous for a new idea sounds good for a startup, but is it? Judging will quickly feel like hurting. So can we overcome?

On top of this, there are the social constructs that led us into our career ages in the first place. As we grew up, some of us might have been taught that as long as we stick to the rules, take the safe road, work hard, and do what we're told, we'll reap the benefits. During our early years, we maybe didn't know we could say no to all this. We didn't know anything about fear or to err on the side of caution. The easy road just seemed faster, simpler. But then again, maybe it led us to a place of slight disappointment. It could be that disappointment, in fact, that has brought you here, considering a startup—specifically in Risk & Safety. If so, thousands of kudos to you.

Listening to one's own heart and soul is a blessing in this world. To get paid for it and have others join you is a miracle. To have a vision and exit with self-worth, ready to start again or retire—well, that better be put into a book, so others can understand and be inspired.

My company fell for it and bought into it all the way, validating itself with a vision of making the world a safer place in a significant and impactful way. The company became my conduit, a second marriage, something that required me to be persistent and not give up. Nothing that had to be done to make it work was a sacrifice; I just did it. Trying to be the best at everything was frustrating, until I learned that all you have to do is just be the best you can be, and if that's not good enough, it's not the end; you just keep trying. Only the universe and God know that rule at the start of the game. For the rest of us, it has to be learned. Startups who stick to the vision and stay strong will be less likely to be knocked off course.

Some will say you don't have the skills. You'll sometimes feel like you have no money, no people, no experience, no time, or you're spending the wrong time, it won't work, or people don't believe in it, can't see it, or prove it. Stay the course. Build and work through it every day. Do this, and your road will be marked with moments that you can be proud of forever. Here are mine:

1994: Started a Safety consulting company
2000: Started a public training company
2002: Built a Safety system certified tool
2008: Built a real-time risk monitoring system

2015: Created an analytics service
2019: Initiated a Public Library of Risk Learnings
2020: Created a data-driven risk template tool
2021: Created a data-driven certification service

Startups see things differently and start with a unique perspective, so it's important to become accepting of learning by doing and taking some risks. Success is a matter of learning through the eyes of the startup, not others.

Regardless of how accepting a marketplace is or how successful you are in the eyes of others, startups can't succeed without failing. Failing and learning go hand in hand. Startups need to fail so they can learn to succeed and exceed.

I was once told by a colleague, "If in the end, it's not good, it's not the end." Startups must therefore embrace the essence of making change, building things, and getting behind great causes. Listening to the "no people" is a sure way to create an end to an unknown. How do I know the "no people" are wrong? Because one will never know an outcome until it happens. And as my colleague pointed out, if the outcome isn't good, then it's not over. A startup's purpose is to find its role in the universe.

Choices

When I first began to explore the idea of a startup in Risk & Safety, one of the most important choices I made was to decide what day would be my last day with my former employer. Of course, this is only the first big choice you will have to make in the early going, and indeed, the need to make choices—and make them well—will never fully go away no matter how long you work or how successful you are in this business.

I didn't recognize this at first, and I wish I had. In fact, it wasn't until recently when I met a life coach that I realized how many choices I had to make over the years and still have to make every day. For years, I'd felt as if a rope was tying me to my desk, but once I realized that this was only an illusion—a perspective that could be easily adjusted—things changed quickly. Over the past four years, the more time away from the office, the more my leadership team was able to take over, the more the daily choices could be shared by that team, and the better my company became. Carrying the torch, finding new sales and clients, building new products and services, hiring new people, and my role...I became more disposable. COVID had a hand in this, of course, because working remotely became the norm. It's tougher to feel tied to your desk when you're working from home. But the point remains salient: as a startup, you will be surrounded by choice; become obsessed with finding new ways to manage those choices and understand that there are always *more choices*.

And if we can step back for a moment, as you're making that first choice about which day will be your last prior to beginning your startup, the second choice to make is about the direction you intend to go. What's your path? How long do you want to be tied to a desk as the leader of this business? In the early days, you'll get to know this business to the point where it will sometimes feel like the only home you need. But is that what you really want forever? What's your perspective on how long you have or decision points to vector off into other things? What's your view on the choices you need to make not just to grow the company, but to grow yourself out of it?

Change and Impact

Consuming the prime of your life while chasing your passion should come with no regrets; you should just do it. I didn't know I would be at this for twenty-eight years (following the eighteen years working for owner-operators like Shell). Building a safer world requires trying things to see what works. This means changing your business frequently and adding new services, people, and products to test and see what makes a difference. Startups need to be looking for the next thing to try, to change and adapt, to see how fast they can grow, how much market they can impact, and how far they can reach into the world.

Some of the changes and adaptations I encountered during my career running this business:

1994 *SIL/PHA/TUV experts*
2000 *SIL tools (SilCore/Profiler)*
2008 *Risk in real-time (Machu Picchu/Sentinel)*
2012 *PHA intelligence reporting*
2015 *Data /Compliance (Risk Alive)*
2017 *Data Learnings Library (100 million learnings)*
2019 *Aggregated Intelligence Platform (AIP)*
2021 *Predictive learning/Laboratory PHAs*
2022 *Brand Intelligence/Self Worth*

As a startup, you're going to be experiencing quite a bit of growth in the years to come. You'll bring in new people, absorb new office space, blow through new computers and technology, and on and on. One way or another, this growth always costs more than money. The other cost of growth is knowing how to embrace and survive the external events and industry downturns that can threaten to derail everything you've built. Whether it's a recession, a 9/11, a bubble bursting in the industry, or a global pandemic, the flexibility to downsize and survive without sales is essential for a startup's ability to survive and thrive in any market condition. Startups can limit their liability with short-term leases and working remotely. The downside is the moving, office setup cost, and employee turnover with frequent change.

Connecting the Dots

Attending conferences was never something I enjoyed, but from the early days of my startup until today, I've done it anyway. Back in 1998, a chance encounter made me realize why this was such an important decision. I struck up a conversation with a gentleman in a small booth way back in the corner of the conference hall. He and I were about the same age, both seasoned veterans in the industry, and with no other foot traffic nearby, I decided to engage him in a conversation and ask what he did for a living. As it turns out, he was a manager for a sizeable east-coast project.

"It's a joint venture with Petro Canada," he said.

With my interest piqued, I asked him if he was experiencing any difficulties in this venture.

He explained that his concern was about finding resources. This was something I could help with. That short discussion led to offices in St John's and years of support and commercialization of my company's first Safety product, developed in relationship with Eastern Canada's offshore projects. If not for that one meeting at that one conference, I wouldn't be where I am today.

As the leader of a startup, when meeting someone for the first time (and no matter what the circumstance), think of it this way: it's your job to meet new people under as many unique circumstances as your startup will allow. If you believe, as I do, that the world is connected, then think of it as your job to connect the dots. Whenever you meet someone, ask yourself, "Why am I meeting this person now, under these circumstances?" There is always a reason, and you and your business will benefit immensely if you take the time to explore that *why*. It may take time to figure it out, but the dots will connect eventually. So when you meet someone, be respectful, inquisitive, genuine, and always on the lookout for that reason why. Do this, and you'll learn quickly that people don't do business with a product or service they need; they do business with *people*.

This is an especially important lesson in the early days of any startup because the hardest thing to do is win substantial work before you have a chance to develop a business record. Working with people is the only way. Without a demonstrable history of doing that specific work, it's going to have to be about trust and relationships. That East Coast contract in 1998 didn't just happen because of striking up a conversation with someone I'd never met before; it also happened because of the colleagues I'd worked with years before. I didn't want to go to that conference, but they talked me into it. The rest is history.

So, be conscious of your most important job as the leader of your new business: meeting and connecting with people.

A Startup Mantra

Next, it's important to have a word, or a few sentences, to meditate on—a mantra. This mantra should include your gratitude, kindness, and intentions as a minimum. Always keep your gratitude to keep in front of you: What are you grateful for every day, every week? At a recent dinner with my grandkids (who I still can't believe are of legal drinking age already), we had whiskey shots to help voice our gratitude, to appreciate the moment and each other. Any time I suggest that staff go for a drink is intended to meet the same purpose. It's time to express gratitude and be with each other.

Finding unique and memorable ways to express your mantra doesn't cost anything. I would sign off emails "With Gratitude," for instance—just a subtle reminder to myself of why I'm doing what I'm doing.

Beyond the reminder element, a startup founder needs a mantra to help him/her disconnect and rethink before every engagement. For instance, "Can the clarity and intentions for the upcoming meeting be stated and validated?" Current assumptions may be untrue; test them always. I tried to ask myself as frequently as possible, "If everything works out, what are my intentions, and are they good for all? If not, what is holding intentions back?"

One of my mantras is "CBOW," which stands for Clarity on truths and assumptions; Best you can be with intention; Opinions by others reflect who they are, not me; Words matter, impeccable and swordlike.

Build Startups on Pilings, Not on Sand

There's an old parable that a house built on sand could float away in a flood. Jim Collins's brilliant book *Good to Great* reinforces the point that we as leaders must make sure that whatever our business does, it has to be about the truth. We must make it real, immovable, hard to argue with. If there are assumptions, we must work to minimize them. If something has already happened, like share price records on the New York Stock Exchange, it's the truth that it already happened. To build upon these truths and make them part of a solid foundation for our business, we must support everything we do. Our products and services remain intact, unmovable, and unchanged, no matter what, supported by pilings driven deep into rock.

For my company, developing those pilings meant finding the truths about what people were saying about us, and just as importantly, whether others were hearing it. If someone says we're good, and that we are safe enough, is that so? It is believable and truthful? In search of the answers, the company leaned heavily on existing data created by clients and owners worldwide, using that data to help them make better decisions on spending and operations (see the graphic below). This was quite the opposite of what others were doing at the time, as they tended to focus entirely on meeting regulatory compliance and legal requirements. Meanwhile, my company focused on the truths, making it more exploratory (and attractive) for a profitable business. And we backed up these truths with evidential information and visualizations of the data.

Considerable Rewards

Yes, the world is connected and has a purpose for you. Knowing your purpose is essential to the goal of making revenue, profits, and sales. It's important to know your change and impact, constantly be on the lookout for opportunities to meet people, recognize your edge, develop your mantra, follow your dream and your heart, and build a strong foundation based on the truth. But you know what else? From time to time, you've got to reward yourself for the work you're doing. You have to take the opportunity to enjoy the ride.

The journey can't be only about constantly striving to improve your understanding and your outcomes and build on your solid foundation. If you're not careful to reward yourself, even if your business achieves the greatest heights, you'll find that it's lonely at the top. You need something just for you, something that's purely *your* reward for doing well. It's about more than pride; one needs a bit of self-assurance that going on your own, starting a company, hiring employees, building, and delivering services/products is all working.

For me, the reward was simple. For some reason—and it's a reason I still don't fully understand—I just need to feel a little prosperous, or at least prosperous enough to purchase a vehicle every so often. So, as my reward for making the world safer, every six years, I would change vehicles. From the outside looking in, no one would have noticed anything unusual about this. This was just between me and myself, a reward that kept me striving through those next six years.

*A **1992** Dodge half-ton, **2006** Ford one ton, **2008** Toyota SUV, **2012** Ford half-ton, **2019** Toyota SUV.*

CHAPTER 8

Risk Is Alive

A Navy SEAL was tossed overboard during a rescue attempt in the middle of a wild ocean storm. The crew watched help-lessly, as there was no way the sailor could stay afloat in the thirty-foot-plus swales. Attempting to reach him would have been a suicide mission for any of the other SEALs aboard the ship. When the drowning sailor disappeared beneath the waves, they assumed he was gone and were of course extremely saddened to lose a fellow SEAL.

When the sea had sufficiently calmed, they sent out a pilot boat to make sure they would not be leaving a fellow officer at sea. Thirty minutes later, one of the crew mem-bers in the pilot boat suddenly spotted something: a body entangled in a fishing line or nets with buoys bobbing in the relentless waves. As they got closer, it looked like the body was alive, with arms waving, and a voice could be heard. Sure enough, it was their lost SEAL. He was alive.

Isn't this just like us? I thought when I first heard that story.

Just like that pilot boat, we in the Digital Risk industry are acting to solve a situation, hoping that with our data we can rescue process units and help them to understand that risk isn't something that remains static. Risk is *alive*.

Unfortunately, hazardous industries have not yet caught up to that line of thinking. Risk, if not measured accurately and precisely, can be siloed and stored away for prolonged periods—five years is the standard, with some facilities not bothering to update until they have an unsafe day.

This leads to a major (and potentially catastrophic) disconnect because the risk profile of a single facility changes, in terms of the sequence of safeguards added, on an ongoing basis. In fact, the risk profile of a single facility changes in real-time as the wider landscape of risk is discovered among other facilities. This is because when one facility learns about changes at another facility, they start thinking more directly about their own risk. And as they think more, they learn more, and their perception of their own threats, safeguards, and impacts changes. If a facility (and/or all similar facilities) faces frequent unsafe days, the perspective of threats, impacts, and safeguards will change significantly.

Given all this, it might be surprising to learn that the average facility has a risk discovery of only 50 percent when first compared to five or more external similar facilities, including the threats linking to hazardous events and the safeguards used. The severities and likelihoods of the impacts can vary just as widely. When you think about all this, it becomes incredibly concerning, the idea that these facilities can go to five years or more before they're assessing their risks. More concerning still is the idea that these risks are not taken and analyzed as risk data.

The story of the rescued Navy SEAL parallels what we do daily, discovering potential storms that could be life-impacting. Risk can be high, fluid, and changing, just like those waves in the storm. And just like the pilot boat, our main character Andrea was recently alerted by one of our team who had genuine concern about the analytics with one of our customers.

The team member had identified a wow factor in a report—an indicator that was almost *weird*. The more he examined the data, the clearer it became that this was a situation with dire consequences if not rescued soon. Think of it like a storm waiting to wreak havoc.

So we sprang into action, alerted the necessary parties, and the crisis was averted. But then, as a company, we were left thinking, *Just because we discovered that warning and saved that site, does that mean our job has ended? Or do we have a bigger purpose here?*

We work with and analyze data daily, understanding that risk is alive, hidden, changing, and waiting for us to discover more. Profiling and finding the risk is our job, but for example, when over-pressure is identified in one area of a process unit with only 5 percent of the known threats (versus the other sixty-one similar units, which show 90 percent more threats), what should we do with our findings? Make one rescue and call it a day? Or does our job need to be continuous?

Are there hundreds of similar sites having the same situation? Is our job informing others of our data-enabled findings and sharing that risk is alive and changing, even if nothing changed at the site under review?

Of course, we now realize, that the answer to all these questions is "yes." Analyzing new data from new sites can quickly change the risk we identify in a given operating site. It's our job to help Risk Management programs behave like risk is alive, help them become data enabled, help them to ingest new threats and events as they are found on comparable sites. It's our job to educate on these matters and change the world of Risk Management forever. Our job is to help equip the world with access to risk data and stories of success and failure.

But this effort to find outliers, inconsistency, and what is critical only matters if the world knows about it.

This realization sent Andrea into planning mode. "Given our daily access and using the data to see what risk really looks like," she said, "I want to take this scenario to our executives and suggest we push the boundaries of our job and build an excellent reputation. We need to share stories about sites that are safe and have fewer surprises. Sites that understand what's critical. We need to convince them that paying attention to this data will allow them to reduce their spending toward the lowest it's been in years. And we can do all this while giving access to our risk data across the full enterprise and each processing site."

The question in her mind was, "If we're working with the data every day, can we hold ourselves to higher responsibility? Can we better serve the world with the knowledge that risk is alive in every process unit?"

It wasn't enough now just to do this within the confines

of our own company. Andrea concluded that it was time to "expand" her team's jobs to where they could share and help the world to see and hear our stories whenever we rescued a processing unit on any given day, merely by finding new data.

Before she approached the executive team, she wanted to discuss the matter with her colleague, John. As it turned out, John loved the idea, but he had some thoughts.

"Can I share one thing I've learned?" he asked.

She motioned for him to continue. "Of course."

"I loved the scenario you shared. When over-pressure is identified in one area of a process unit with only five percent of the known threats, while the other sixty-one similar units show ninety percent more threats, what should we do with our findings? I have a few examples of what I'm learning about the data on any of the 250 processing technologies ingested so far." John smiled. "You know me; I'm a bit of a nerd. But I'm also the one who builds the apps in the background to help the experts, and every team member of a PHA team expands their curiosity on the world's largest risk database."

Andrea approved, giving John the nod to continue.

"Being curious one day on this unit, I asked across our enterprise of process units worldwide: 'What is the average ratio of threats to safeguards, looking for outliers and highs and lows?' I found three or four other units—different owners, yes, but they all had the same situation of low threat counts on similar process units. I did see the same results across the enterprise in a few others, but then, being the nerd, I asked, 'What are the top safeguard types normally deployed?' And then I also asked, 'What is the average threat

THE SOUL OF RISK

count used for each safeguard type?' And again, the findings across our enterprise of units found the same outliers of three or four."

This was all Andrea needed to hear. "We have a higher level of responsibility to let these other facilities know what we learned," she said. "They're having the same issue as we've had. We can help them prevent an accident."

"We don't know exactly who they are," John warned, "but we do know the facilitators of this enterprise data, so maybe they can track them down and let them know what we accidentally found."

Herein lay the hurdle. There is a higher purpose, and yet there is work to be done to deliver the data to the people who need it.

"This discussion is bigger than just us," Andrea said. She thanked John for his input, then immediately raced to set up a meeting with the executive board.

A Watershed Moment

Whenever we have an unsafe day in the process industry, the real learning tends to stop. The race is on to find something to say in response to the events of that day. And often, the pertinent information, or the real reason for the unsafe day, can be lost within the tangle of appeasing insurers and lawyers. This, of course, is the opposite of what should happen.

Data is an investment, and it should be considered an investment on the level of building safeguards, operating procedures, and funding inspection programs. If we're going to be spending consistently on Risk Management anyway, then why not think of data as a safeguard and connector to everything we do? The data may originate from our PHAs and HAZOPs, but analyzing and learning from the data, and deploying the learnings and the benefits across the organization, is a much bigger and much more beneficial effort. The time has come to improve our ability to see what clearly matters. Data is not a one-and-done thing; when used properly, it allows us to measure risk and take steps toward ensuring safety in real-time.

Even though risk is alive, fluid, and changing, access to the data allows us to see it with greater clarity, in a timelier fashion, and in an incredibly effective way. With all the ever-changing data from MOCs and PHAs, we get to compare ourselves against others.

Anyone in the industry who thinks that this future we're discussing here is only an extension to PHAs is thinking too small. Data has become an everyday part of everyone's lives. In the world of Risk Management, the data starts to influence our ethics. It allows us to question whether we're doing enough to keep ourselves safe. It gives us the means and opportunity to ask new questions that will help us improve across the board.

Risk management has always been a critical aspect of any industry, but with the power of data, we can now see it with greater clarity and make informed decisions quicker than ever before. The insights we gain from data are changing our culture and influencing our ethics, and it has become clear that we can no longer assume we're safe without it. Data has become an essential ingredient to our lives, enabling us to make better decisions in everything from health to climate to risk management. And as we continue to invest in data, we can shrink budgets in other areas and truly change our culture, leading quickly to profits. It's a measurable piece of our business that we can no longer ignore.

Looking forward, we as an industry must see data as an investment into safety and reputation, something that helps us define who we are. We are coming to a watershed moment that, like PCs, like the internet, like smartphones, and like AI, will change the world forever. We simply need to work together, across industries and facilities, to make it happen.

CHAPTER 9

Risk & Safety Is Good for All

"WE ALL KNOW that focusing on Risk & Safety is good for all. So I guess my big question is, why is it so untouchable? And why does the existing process take so long?"

Andrea looked around the executive boardroom at the rest of us gathered at the table. The silence she received in reply was making us all feel a little uneasy. Sure, as a company, we'd been experiencing the effects of the Great Resignation just like everyone else, with a wave of voluntary turnover costing us valuable talent and knowledge, but we hadn't seen an unsafe day or even a near-miss for at least two years. Our costs were escalating, and likely it was only a matter of time before our decision to cut back on training costs and external PSM would lead to a near-miss or unsafe day or two. But Andrea was a young hire, new to the risk industry, and this was her first meeting with the full board. She'd graduated at the top of her class, we'd hired her even when we weren't exactly looking for anyone to fill the role, and she'd quickly proven herself worthy to lead a growing team—but still, who was she to ask such big questions?

"Thanks for your questions, Andrea," I said. "Please forgive our startled silence. We'd hoped that you being from a new generation would bring a fresh perspective and help us better understand how to be safe."

"Well, I'm glad to hear that," Andrea said as she wrote something into her notebook. She sat back and looked directly at me. "But it doesn't really answer my questions."

I broke into a broad smile. "We've been following our PSM thirteen-element checkbox program, which you're right, does take time. Truthfully, it's also costly, requires lots of effort, and the gains it produces are hard to measure."

"All the more reason for change."

"And this is exactly why we brought you in, Andrea. We've kept safe even through high staff turnover, COVID, and so many of our key people working remotely. All that said, we've seen the near-misses and process incidents at other plants, so we recognize that we could be stunned by an unsafe day at some point. We can talk process all you want, but a huge part of what we need to do in the near-term is to ensure that our sites are staffed to where they can operate safely."

At this, Andrea began the presentation she had prepared for the meeting. My eyes grew wider and wider as she progressed through the slides, each of them asking salient questions that none of us in my generation had ever asked before—questions like:

- Can we utilize advanced algorithms or AI to better predict incidents?
- Have we asked about everything that could go wrong? Are we sure we haven't missed anything?
- What are others predicting and considering outside of our facilities, our organization, and our industry?
- Are we inconsistent in the way risk is viewed between sites? If so, how can we quickly and easily identify these inconsistencies?
- Is there a consistency of the severity and likelihood values that we assign or are assigned by others?
- Are we happy with efficiencies, culture, and growing spend on Risk & Safety? Are we spending on the right things, the optimum safeguards, and the best risk reduction measures? Or are we spending on the wrong things, the zero risk-reducing safeguards, and flawed recommendations/actions?

In systematic fashion, Andrea's incredibly insightful presentation took us through the subjects of risk discovery, health checks, ROI checks, and our Aggregated Intelligence Platform (AIP). When she was finished, I cleared my throat and asked if she could summarize the goals she hoped to meet.

"I want to help align this company with the most modern resources," she said. "My team's research is telling us that our decision-making process is lacking the perspective and the thinking of the young."

"Are you suggesting that we're all too old and out of touch?" I asked with a wry smile.

"I wouldn't put it that way exactly."

Everyone in the room chuckled.

"Your point is well made," I said. "Being out of touch with technology and changing perspectives, well, it sneaks up on you. We have no intention of ignoring the ways of your generation. We're all very grateful for your input."

"I'm glad to hear that, because I think with the technology, along with my generation's perspective, we can make this site and all sites much safer—and in a far more efficient and cost-effective way."

"Wonderful. Enlighten us on how you intend to do that."

With that, Andrea launched into her description of the revolutionary GRIN system. "We're going to leverage the data now available across companies, industries, and in fact, the whole world. Instead of just relying on our own data, we're going to use the aggregated intelligence of others. This data will provide more just-in-time Risk & Safety intelligence and decision making to everyone across the org. I'm not just talking about us. I'm talking about site operators, maintenance, engineers—everyone will have access to a data-driven platform that will enhance safety. This will help us modernize and leapfrog our work processes and tools to enable future generations to do everything better."

The stunned silence had returned. A few months back, we had wanted to hire a smart, capable young person to show us how we could utilize modern technology and data to improve, and it looked like we had gotten exactly who and what we'd hoped for.

But the silence wasn't just about how impressed we were with Andrea and her team's findings; it was about how our eyes were suddenly open to possibilities we had never considered before. This was not the first time we had questioned what the organization was doing. Our board and our CEO wanted to know what could be changed, and we'd asked many questions about digital transformation and the focus on Risk & Safety. The problem: safety wasn't scheduled to be reviewed—at least not in quite this much detail—until the next quarter. Kudos to Andrea for her questions and comments. In so many words, she was suggesting that we could access the world's intelligence to make ourselves safer. She clued us in to a key fact that far too many people in our seats continue to overlook: the world is now connected, and we have a responsibility, if not an obligation, to use that connection to make ourselves and everyone else better at what we do.

"It has become common to utilize technology to predict the future," Andrea explained. "That might sound like science fiction, but it's true. Almost all things can be measured and predicted as long as we have enough data, analytics, and technology. So even though we might be carving a new path here, it's likely that predictive modeling will be useful for Risk & Safety as well."

"What do you need?" our executive VP of operations asked.

Andrea suggested that we give our collective data of threats and impacts to the younger members of the staff, and that they explore it quickly, hoping to find a new way of seeing Risk & Safety. She suggested that we use the space and time of every safe day like a blessing—one that helps us set goals and continuously get better at predicting and seeing the risk before the next unsafe day.

As some among the executive team began to speak about next steps, the rest of us were left wondering if this was our new reality, leaning on the young for their new way of thinking about data and intelligence gathering. Many of the questions Andrea had posed were questions we hadn't openly asked ourselves yet. But maybe we just didn't have the perspective necessary to ask the questions in the first place.

"Do you see where this is going?" Andrea asked.

"How do we introduce this material into this organization?" our manager asked.

"Exactly," Andrea replied. "The plans I've outlined can be perceived as disruptive and new."

"We need to change our perception on the importance of this," I chimed in.

"And how do you recommend we do this?"

"I think it's best if we position this plan as a part of the effort to keep our people and leverage the Great Resignation to retool. We can frame this as putting safety first, making use of data that's already on the shelf and paid for."

Andrea picked up on this notion with a burst of enthusiasm. "To avoid a lot of work, gathering data, creating something big, we could assign two technical people to the task."

"Andrea?" our manager said. "Can you give us some insight as to why you believe a couple of young technical people, first career or second career, will find something to help us?"

"I'm confident they'll find something. Someone must be doing this already. I'll ask them to find vendors, tools, and data. In some cases, it might be better to simply purchase the data and learnings. The purpose will be to visit the data findings and look for plants like ours. I think we can share our data, keep the source data secure, and have someone with the expertise and tools to mine our data and extract the learnings we're looking for. We don't have to share our source data or spend any effort ourselves to do this. Our first priority will be the investigation into what data services are available and put into practice around Risk & Safety. Let's start there."

"I hope you're right," our manager said. "I've never seen or heard about these data services specializing in Risk & Safety."

"Have you ever done a Google search on something?" Andrea asked wryly, and to a chorus of approving laughter. "Well, we can start there. And we'll find data-related links to plants similar to ours. We'll also find products that offer intelligence from other datasets on sites like ours. Here, see this slide..."

With that, Andrea clicked to the next slide on her presentation.

Industry Analytics on SISs, Safety Instrumented System

Industry: Petroleum Refining
Processing Technology: Fluid Catalytic Cracking (FCC)

How many SISs were found in similar refinery process units? Why does this matter?

Safety Instrumented Systems (SIS) respond to developing hazardous events in processing units by preventing or mitigating the potential harm or damaging consequences. They are also associated with HIGH CAPITAL costs and must be maintained, proof tested, to retain their intended Safety Integrity Level (SIL).

Industry Data

Are the number of installed SIS in your process unit in-line with industry best practices?

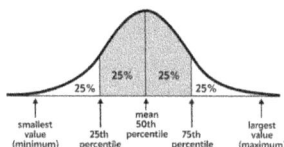

smallest value (minimum) | 25th percentile | mean 50th percentile | 75th percentile | largest value (maximum)

How do I compare?

Are you in the bottom quartile?	**Are you close to the mean?**	**Are you in the top quartile?**
Then you may have missed some high-risk scenarios typically requiring a SIS. If you missed high risk scenarios, you may be exposed to a potential hazard occurring.	Then you are within industry norms and your process design is in-line with recognized industry practice. Your knowledge of the process is good, and your risk perception conforms with industry. Risk neutral.	Then you may have perceived scenarios that are perhaps not considered to be of high risk. You may have used too many SISs to close the risk gap unnecessarily.

Related Facts

- 95% of SIS safeguards in FCC units are reducing the likelihood of hazards with "HIGH SEVERITY" consequences
- 1% of SIS safeguards in FCC units are reducing the likelihood of hazards with "LOW SEVERITY" consequences
- SIS Safeguards represent 5% of the total unique safeguard but contribute 80% of all high-risk reduction in FCC units

Now What?

Look at your most recent FCC PHA and count the number of SIS safeguards to determine how you compare to industy reflecting best practices.

Learn More

What is SIS?
Functional Safety Primer

Fig 1.0 Sample Question answered with Risk & Safety Intelligence.

"Like magic, this vendor has created data packs dedicated to hundreds of commonly asked questions, the answers drawn from a database ingested, mined, and built to ask questions and learn from them. This is the kind of thing we are looking to explore and then determine our strategy and effort involved. Maybe we just need a point person to manage the process, and a vendor to monitor the flow of data. And then another to help integrate or leverage the learnings. I'm guessing that we want to improve our efficiency, reduce our spending, and make it easier in the future."

"Brilliant," the manager said. Hiring a first and second career person makes sense to help us get going. And under your guidance, Andrea, right?"

"Of course!" Now her smile faded as she returned to specifics. "It's important for us to pilot everything we do. We must build the case if we want to promote it as an enterprise solution. So let's be ready to share the results with other business units."

All the heads in the room were nodding.

"I'm predicting we'll succeed in this," Andrea continued. "We want this path to be business-driven, a path that makes sense not only for our site, but potentially all sites. Limiting the work and cost to two new hires and a vendor to assist us through a pilot, we can justify and get behind this. We do this with the intention to significantly minimize costs, find work efficiencies, and be smarter. That has a long-term operational savings impact, year over year."

"This should play out well," I chimed in. "Staff is being turned over for so many reasons."

"The technical part is the easy part," Andrea replied. "My

grandpa, having his own business, taught me that making significant business decisions and seeking value and ROI from technical choices is essential. Always look to minimize the effort and cost to understand the importance and intentions of all decisions. Making decisions on technical merit alone is not good business."

"I like this new hire," one of the elder board members said.

A rumble of laughter followed.

"It's like she understands our predicament, our situation, our fears," our manager said.

When the CEO asked for closing comments and questions, the room kept silent, maybe to avoid sounding like we knew that what we'd been doing was no longer good enough and would have to change. None of us wanted to dwell on the past or dig our heels in and keep doing things the way they'd always been done. We wanted to move forward and explore the question of what else could we do to be safer, more efficient, and more profitable into the future?

"Let's handle this path forward with a steering committee of operations, maintenance, HR, admin, engineering, and HSE," the CEO was saying. "I'll be part of the upcoming meetings and want to embrace and help where I can." Then he set both of his hands on the table and looked around the room. "My eyes were opened today. This is not some futuristic event. We begin a new journey today. Next week, let's have Andrea demonstrate some of these questions and comments to increase our understanding and bring in the right people to make this happen. It's a journey, a change, not a project. Let's be influencers and key contributors, and let's let our young and integrated teams do the work and set the path."

In the end, the board came together on the plan with enthusiasm. The way Andrea framed her case, just two short weeks later, the new people were hired. The big hurdle, having the organization consider a new way of looking at things in order to see better results, was refreshing. In our world, Risk & Safety had not kept up with technological, cultural, and professional change. Now was the perfect time to hire, leverage new perspectives, give the new hires permission to do and search without any past or biases weighing them down. The opportunity to grow as an organization through a renewed approach to Risk & Safety could not have been brighter.

It felt like our organization was leading, and attracting, and setting new strategies, and maybe becoming the envy of other companies. We were acting alone, but we needed to do something different. Budgets, unsafe days, and staffing levels continued to grow. It was time for a change. We made that change. Now, the future could not look brighter.

And what is the best part? We are only one organization. There's a whole world full of others just like ours, and we're ready to help them make that same positive change, for the better of all.

CHAPTER 10

The Evolution of an Industry

Imagine that you are the one person on the planet who is responsible for all the risk in the whole world. Imagine waking up in the morning and checking your messages, finding out that you are on the receiving end of every unsafe day, every unhappy employee who was harmed, every member of the public who is concerned about the environment, and every company worried about its reputation. Just a *lot* of people who are extremely upset with you.

They're upset with you because you didn't warn them, you didn't predict that something harmful was about to happen. Isn't that why you're paid? Isn't that your job? Well, you're fired! We're going to replace you. We can't tolerate ammonia leaks, runaway exothermic reactions, fires, or explosions. It's bad for business.

Think about this for a second: Why are there so many job openings in Risk & Safety? Maybe it's because one must have broad shoulders and thick skin. You can either take your day-to-day career as a call to just do the same thing the last person did. After all, if you don't change anything, who could blame you if things go wrong? What the industry needs is more people who think bigger. We're just waiting for someone like you to change, fix, and try new things.

In the years to come, the number of Risk & Safety jobs will continue to grow across all sectors, especially in Data and Digital Transformation. And it looks like insurance costs will continue to rise, so jobs that involve holding the line or reducing insurance costs will make for a good and stable bet to turn into a long and rewarding career. This is especially true in sectors related to healthcare, oil and gas, manufacturing, government, and IT.

The outlook for jobs in Risk & Safety is promising, but at the same time, the industry is currently extremely fragmented. The good news is that there are multiple startup companies aimed at serving the underserved Risk & Safety market in ways similar to how Uber and Airbnb served the taxi and hotel industries.[7] These efforts will radically increase the number of jobs and people working in the Risk & Safety sector, because each new hire can work as much as they want, and work with tools and platforms that are easy to use from home.

---

7   I write this with an important caveat: Uber and Airbnb certainly aren't flawless companies. I make the comparison here only in the sense that I'm envisioning a similar leveraging of technology that will completely change the way the risk & safety industry operates.

And it's not only startups who need people. Insurance companies will be looking to hire data analysts, and people with backgrounds in Risk & Safety because they also see that the world wants others to benefit from the data we can glean from unsafe days. The more forward-thinking insurers recognize that the old days are rapidly ending—the days when insurers, lawyers, and owners work behind closed doors, hiding the data and controlling the timing of learnings released from unsafe days, often ten years after the fact or longer. The more the data is readily and openly shared, the more jobs working on developing Risk & Safety software platforms there will be, the more the industry will need auditors, investigators, risk analysts, data analysts, and on and on.

Every time there is an unsafe day, these new companies will need your help to collect all the truths to uncover the threats and impacts and convert that knowledge into data that can be shared with the world immediately. Your job, if you choose to accept it, is to use the data to change the past work practices in Risk & Safety and provide for a safer future.

Early Days: Send Us Your Data

The company I founded, Risk Alive, had a tagline in the early days of pioneering the use of data: "Send us your data, and then we'll have something to talk about." The idea was to bring awareness to unused data, and it fell right in line with the digital transformation trend.

Back in 2019, Risk Alive used this message with a pro-spective client in the energy sector, and they agreed to let us have a look. Just as we expected, but no different than many others, they had potential weaknesses. They were missing threats and safeguards, and they weren't recognizing unidentified hazard events that had been considered elsewhere in the world. No excuses here, but like every other owner-operator we approached, until we came along, no one had thought about comparing site-to-site data. Most people didn't even think it was possible.

At the time, we knew immediately that we would find and uncover risk that the prospective client hadn't considered. They expressed a familiar brand of skepticism, to which we had learned quickly to respond, "Hey, that's my job. I got this. I'll just need to review the findings to ensure it's accurate."

We had spent the past seven years building software to help us do exactly this. When we came back to them, we shared over a hundred years of related data experience by similar facilities. It was so disruptive and different and new, it seemed as if the decision makers found it difficult to accept something that had processed the data so well.

Unfortunately, the old brand of thinking prevailed: *It's too difficult; garbage in is garbage out; it's impossible to predict these events.* And like many others, this prospective client elected to ignore the data learnings.

Later that same year, they experienced a catastrophically unsafe day. A pipe elbow corroded and failed, to the point that it ruptured and caused a massive release of toxic acid, which in turn triggered massive fires and explosions. The estimated property damages approached a billion dollars. The number of people potentially impacted by the toxins released into the atmosphere and water supply was in excess of a hundred thousand.

Soon after, the business went under. We weren't sure if we could have prevented what happened, but the event validated why the world needed to know how we were using data to help uncover hidden risk.

Risk knows no boundaries. Risk doesn't care where the weakness is; it only knows the errors, threats, impacts, hazard events, and safeguards, regardless of the industry. So another heads up, for those seeking careers in Risk & Safety: it applies across all industries. We have had employees that started with us in the energy industry, and others that began in the food business. Still others were in the automotive safety business. Wherever you come from, the risk methods are similar and highly transferrable across industries, giving those in Risk & Safety careers a lot of choice and runway.

Anyone looking to enter this industry needs to be able to communicate findings in a business sense as well as a practical sense—messaging that will resonate with the people doing the work on site as well as the ones setting the budget. Running analytics, generating interesting findings, and sharing opportunities to reduce cost and make better decisions across organizations—all of these efforts need the right words and messages.

And let's not forget the insurers! Sure, they are the only ones to benefit from the threat of unsafe days, but they also need to (at least appear to) understand everything about risk. Who better to offer their extensive accident databases and consulting services than insurers? So the growth in demand for Risk & Safety people with a data background will be tremendous in the insurance game too.

Careers, in this new business of data-driven Risk & Safety, although cool, interesting, and exciting, also have a lot of challenges. Namely, with 250,000 sites worldwide, what could possibly go wrong? Risk doesn't care if it's an HF Alkylation unit, a refueling station, or an LNG facility. Risk can spring up based on errors, threats, impacts, hazard events, and unavailability of safeguards, including procedures and inspections. This means that risk is seeded everywhere and grows everywhere, as it's watered with time, with changes to the process, and with a lack of awareness. The only thing is for each owner to implement a full risk management program. It is this author's opinion that anyone with a data background who starts careers in Risk & Safety today will soon be running those risk management programs.

CHAPTER 11

The Breakthrough

One day, Flynn was looking to hire his first VP of Risk Excellence. The current direction was expensive and not scalable. Somehow, Flynn felt he needed a new perspective, a breakthrough when showing and speaking about Risk Excellence, someone who could demonstrate that this was disruptive and new. Flynn saw these experts as underpaid and overworked, and he knew they were likely frustrated about having to compete for work based on price rather than the value of reducing the likelihood of unsafe days, increasing profitability, and creating spending efficiency.

Put simply, Flynn wanted to get back to the basics of business. He wanted to find a long-term, sustainable way to reduce large amounts of risk for the least effort and cost, with the most intelligence, and based on the data and the evidence. He knew that there was so much more that could be done to be safer, but first, he would have to overcome the tight budgets and the checkbox mentality of getting risk assessments and recommendations done just to keep up with compliance.

Flynn's new way was more accurate and efficient. If only others could see this and be open to considering a new approach. Hence, he needed some serious help from a VP with solid credentials and the kind of speaking ability that can help stubborn people see new realities. But he had spent weeks discovering the hard way that people who were qualified for the role just couldn't seem to get past their established thinking on how Risk & Safety is supposed to *work*.

By the time of his third interview with Darlene, who was currently working for a prominent risk certification and consulting company, Flynn was getting a little desperate. It also happened that Darlene was the perfect candidate, the best he had interviewed yet. So instead of running a standard interview, he decided that it was time to make his case to Darlene with a formal presentation.

Through the review phase of the presentation, it seemed that Darlene was interested, but everything she said made it seem as if this wasn't really for her.

That's when Flynn turned a corner by showing Darlene the cost of each safeguard and making a more pronounced case about the business relationship between risk reduction and fees.

"Specific recommendations can cost either less or more," he explained. "Some achieve high-risk reduction levels, and some achieve near zero risk reduction."

Then the second function in the demonstration started. Flynn made clear the case that the overall strength of the recommendations depended on the *sequence* of the recommendations. Each sequence had a varied impact on the overall risk of the facility. Being a perfect sequence, one series seemed to reduce the risk quickly, and after only a few recommendations were applied, the rest of the recommendations seemed redundant. How could this possibly be a sound way of doing things?

Next, he showed how, with all of these things being true, many recommendations and related expenses were not contributing to the risk reduction in any significant way.

"Oh, wait a minute!" Darlene said suddenly, and with wide eyes and a raised voice. "We don't do any of this. I've never seen anything like it at my current employer."

"So how do you assure your clients that their spend is going to the right processes?" Flynn asked with a knowing smile.

"Our clients simply trust us," Darlene said with a shrug. "We call it a Gap Assessment. We go in with five auditors for a week. And those Auditors pick a hundred or so safeguards to audit out of the eight thousand or so safeguards on the site, including procedures and inspections, in addition to some that might be alarms and operator response, safety instrumented system trips, or mechanical safeguards like PSVs. The audits randomly check for the existence, proof of availability and operation, maintenance, and testing. If certain areas in the process are sampled, searching for gaps, then further studies or actions are recommended to close."

"That's pretty standard," Flynn said, gesturing subtly at the not-so-standard solution he was demonstrating on the screen.

"We don't have what you're showing here though," Darlene said excitedly. "We have no way to measure the complexity, criticality, and cost and then rank all of it to help us see and understand what matters more. We have access to client information and their past risk assessments, but we've never combined it with other data to see a holistic view of the risk. I mean, I can see how this would show us that we're missing safeguards, threats, and impacts that other sites have experienced and considered."

Now it was clear that Flynn had Darlene's attention. Unfortunately, their time had run out. But what happened next was pivotal—an unexpected game-changer that Flynn hadn't even dreamed about. He had arrived in the hopes of hiring someone who was willing to buy into the vision of the data, someone who was full of energy and experienced enough to convert others along the conversion line, someone who stood out as a representative of Risk Excellence.

But then, suddenly, Darlene said, "I'm going to send you some data from my current client. They're large and ambitious to do the right thing about process safety in all related facilities."

"How large?" Flynn asked, his heart pounding.

"They control all their home country's facilities, from power to fertilizer to Petrochemical."

It was all Flynn could do to keep his eyebrows from flying off his head.

"I see how what you're doing is disruptive and wildly different," Darlene said. "For every site manager to be able to quickly, easily, and accurately see through the complexities of what could happen? That would be an incredible blessing.

And gathering this information from others operating similar sites makes so much sense. One database of learnings gathered from the world experience becomes a foundation for everything we determine to be a risk, everything we spend effort and time on, not even thinking about the costs we likely misspend today in the name of being safe. An owner's dream of a hazardous facility is to know that they've done their best on a world scale to do the right thing and protect their employees."

"Exactly," Flynn said. "I'm so relieved to hear you say this."

Two days later, the data was received, and Flynn started working on it and finding the intelligence about being safe and the business just staring us in the face.

In summary, what Flynn learned from Darlene that day was pivotal for both. The intentions of each, with no agenda, of selling but serving the idea to discover something new and determine its value. For Flynn, he had been struggling to find a repeatable process to increase the slope of the conversion line about the well-known craft of Process Safety. Now, there was a light shining on a path, one so bright that others could see that following the data could have the same impact on their organizations, without suffering, damaging reputations, or incurring tremendous cost. For both Flynn and Darlene, their lives would change forever.

Darlene started working for Flynn's company and making a difference in ways far more exciting and rewarding than he was initially considering. Leading up to that day, he was essentially just serving as an instructor, teaching others what many others were already teaching, or conducting risk assessments that thousands of others were already conducting.

This new path of following the data was so much more rewarding. Flynn never looked back. He tasted the secret sauce, so others could be introduced to this new and disruptive approach to following the data. He and Darlene kept themselves in a neutral space in meetings—never selling or promoting, just talking and telling stories about the impact of what they were doing, everything coming from evidence and truths that could be demonstrated. They targeted those who seemed most ready for a change, most ready to make a difference, no matter where they resided in the world.

It didn't take long for Flynn to realize that what he needed now was even more help.

The Author's View of Risk & Safety Entrepreneurship

Just like Flynn, I have served as an entrepreneur in the Risk & Safety industry, and I've been doing this long enough that I have picked up a few pieces of wisdom along the way.

First, nothing great ever happens without taking a chance. Entrepreneurs, passionate about making a difference, having purpose and impacting an underserved marketplace, can find everything they need to turn their dreams into a reality. However, the journey is not easy, particularly in this business. It is filled with countless hours of work and difficult choices. Taking that first step and embarking on the journey to open a business is definitely a brave endeavor.

But as mentioned in the introduction, the intention of sharing the journey of the company I founded is to inspire others. The hope is that others won't have to live through everything we lived through to become strategic, make better decisions, and focus on the more relevant decisions to our business and its people.

After so many years, one learns that it is an honor to lead and gain the trust of clients, employees, and others to follow your lead and vision. Even when they didn't understand everything but believed it would be worthwhile, being part of something as big as the dream felt like the right thing to do at the time.

The beautiful thing about taking a risk and starting a new venture is not knowing exactly where it will significantly impact those who participate. For this company, it started with building automated safety systems known as Safety Instrumented Systems (SIS) and complex and highly available safeguards designed and tested according to strict specifications. This was, like with most startups, a general direction, an underpinning idea. With some expertise, and enough knowledge to build a craft and repeat it as a product or service, maybe it will be successful.

This business of helping others can be challenging, and it's not for everyone. It's a journey. It's years. It's sixty-plus hours a week. It's travel. It's seizing the opportunity to make a sale, and it's doing whatever it takes. Hanging up a shingle and saying "expert for hire" is one thing, but employing others, mentoring others, stepping aside to let others do what you took pride in, and hiring others to do what you never learned to do, is a whole other world.

It has taken me a lifetime trying to find a company that would revolutionize something and attract employees and customers. Once found, the challenge is letting the world know that having tools to share and access the data is essential and beneficial for owner-operators who want to be more efficient, make expert decisions, and make good decisions without the experts. Internally, the struggle is to help employees understand the short-term and long-term goals and accept change as a good thing.

Risk & Safety Management is a complex field, and the process of tracking and managing risk data across hazardous sites is often slow and tedious. By utilizing a cloud-based platform, organizations can take advantage of rich data management, analytics, and intelligence capabilities to help them get the information they need quickly and securely—from preventive measures to prospective changes to proactive decisions. This helps organizations minimize costly damages resulting from unforeseen risks and truly maximize the benefits of cloud-based tools.

Taking these steps will allow organizations to remain at the forefront of the Risk & Safety Management field. It will allow them to scale their product to the world just like Netflix once did, starting with mailing DVDs to customers and eventually growing into a global phenomenon of a streaming service. Along the way, Netflix has retained a significant percentage of its employees with competitive salaries, flexible work schedules, and generous stock-based incentives. They are also well-funded, having raised over $1 billion in funding from investors over the years.

Or consider Airbnb, which in a matter of fifteen short years, has become an international marketplace for users to find or list lodging and other rental properties. They got there in similar fashion: offering their employees attractive salaries, stock options, and other financial incentives to help retain talent. Their funding from investors has allowed them to grow their platform to over 220 countries and territories worldwide.

A similar wave of growth is just waiting for us in the Risk & Safety business. Every piece of intelligence we collect is worth something to the right organization and the right situation. Imagine what thousands of pieces of intelligence can do for the world. That same kind of imagining has grown my company's brand considerably over the past seven years. For you, just starting out as an entrepreneur in this industry, marketing to and connecting with a connected world has never been easier.

I started down this road when I was 64, having claimed that life is short and that I needed to maximize the value to all shareholders. Now, four years later and at the age of 68, it's only become more evident how we do that. We got to where we are by putting a successful team together. Based on their observation, clients needed our help to do more than just extract the necessary intelligence; they need to be shown how to deploy and use it. While clients stuck to their core business, we helped them become efficient, improve their quality, and reduce their spending. Your company (or soon-to-be-launched company) can do the same, ensuring that the client takes full advantage of what they're paying you for.

In this exciting and rapidly growing industry, I wish the same success and more for your own startup business.

CHAPTER 12

Conclusion: The Future Needs Our Data

WITHOUT AN EFFORT to compile and share the data that hazardous sites have collectively gathered, everything else we have learned over the past forty years may not survive. Meanwhile, the craft as we know it might not be recognizable in the future. But the data itself has survivability written all over it.

The younger generations learn in different ways, are comfortable with scripting without the cost of pretty interfaces, and are extremely resourceful in a Google- and AI-driven world. It's a given that if they are presented access to the data, they can figure out how to use it to make all sites safer in the future.

Risk Alive, the data-enabled risk consulting company I founded, is only one company—a living, breathing example of what the future can be. It's the first of many, with a new recognition that risk is alive and always changing, and that the old, prideful thinking that "unsafe days don't happen here" is replaced with humility and curiosity. The transformation is toward a new line of thinking: "We don't like spending on the wrong things or committing too much funding to the most minor risk-reduction efforts." Or, "We don't want to discover that the money we spent over the past ten years was misspent on low-impact or ineffective actions, equipment, and training." Or, "It's important to know that going forward, we can see risks more clearly."

We're dismantling the traditional ways of thinking about "complete" Risk identification and control. If events, threats, and impacts identified by others remain unknown to you, then how can we say Risk is "completely identified and controlled?" We're changing all of this by replacing traditional tools with learnable, leverageable information that can be accessed and implemented in seconds.

In the meantime, we're on a journey of always finding exceptional value for each group. Some are finding profitability by seeing what matters most. Some are finding efficiency and knowing what to work on first. Some are finding new careers, and it's exciting, engaging, and attractive.

Change (and Savings) without Accident

Risk is alive and always changing. If it doesn't change, it's because there has been an accident of some kind. Following world trends toward analyzing data, and evolving Risk & Safety to being data-driven, means that learning and change can occur all the time without the pain and suffering of unsafe days. And it can be more predictive and proactive versus reactive.

After implementing a data-driven model, the most fun and significant part of the first few years is predicting *savings*. Making informed decisions across an org with wide access to all the data and learnings leads to lifecycle savings and puts site people in a position of strength to be safe beyond words. This is because, suddenly, the people on site have the information they need on a timely basis, information that helps them make the right decisions, even if they don't have expert status. The data is incredibly empowering.

Who Benefits from Data as a Safeguard?

We're currently facing a problem in the Process Industry. It's clear that the owner-operators of hazardous sites handling or producing the inherently dangerous product(s)—and the unsafe days impacting employees, the public, and the environment—are the primary beneficiary of using data as a safeguard, but they're also not necessarily willing to be the owner of the problem. So, who is the owner of turning data into a safeguard?

It is also clear that there are secondary or indirectly contributing groups who could choose to demonstrate and promote data as a safeguard. And they have been doing so in recent years. American Fuel & Petrochemical Manufacturers (AFPM), with its hundreds of refineries in North America, is an example of an organization that chose data as a safeguard for the overall good of representing common interests. AFPM believed in the data, and recognized what it could do for the member companies and their own Risk & Safety. Although they are not directly benefiting, they are promoting the good of using data as a safeguard and sharing the benefits directly with their members.

Other secondary groups (and indirectly contributing groups) could choose to do the same thing, promoting or demonstrating the use of data as a safeguard. The manufacturers of existing safeguards (e.g. Honeywell, Siemens), Safety Instrumented Systems (SIS), alarms and trips (BPCS systems), and fire and gas detection systems could choose to promote and demonstrate the use of the data collected by their own systems, or they could choose to benefit directly by understanding existing risk data as a safeguard.

Still other secondary groups are in similar situations, like the insurers, auditors, regulators, public interest groups, who could decide to use their own information and reports as a data source and share it with the world. Or they could promote and demonstrate the risk data as a safeguard. For example, an insurer might decide to leverage all their global data, and share all PHAs, HAZOPs, etc. for the good of data as a safeguard. Or they could leverage owner-operator data to help demonstrate the good of seeing their data as a safeguard. Either choice is better than just reporting losses and accidents. Either is proactive and change-impacting. Either would serve for the good of all industries, reducing risk and sharing the benefit of reducing risk among any of those interested.

And that's not all! Still other secondary groups are in the same situation. Auditors, procedure authors, maintenance vendors, licensing and processing equipment manufacturers—all of them are holders of information not currently being used as data.

The Path Forward

So is no one doing anything to solve this problem? Has there really been no progress toward this brighter future?

Some among the primary beneficiary groups, owner-operators of sites generating and handling the hazards, have gone through pilots pushing the use of data as a safeguard. I am unaware of any that have stopped this effort and elected to go back to more traditional ways of seeing risk or operating without data as a safeguard. Those groups are benefitting directly from risk data as a safeguard and have given rise to the notion (and proven it, in most cases) that using data as a safeguard is safer and increases efficiency, quality, and profitability.

None of these primary groups felt they were responsible for delivery of data as a safeguard, but all were interested in seeing the results of applying it. They all found that analyzing the data, ingesting the data, collecting the data, and aggregating the learnings lead to Risk & Safety measures becoming the cool new thing to do. They found themselves re-engaging and being revitalized with technology. It inspired a new craze of thinking, producing visuals of what the data is, putting giant risk metric cards in meeting places, control rooms, lunchrooms, and chat rooms. Soon, operators and site people had learned about the top three threats, top three safeguards, and top hazard events in every process unit they operated. Using their contributions and thinking power, they figured out what this information could be turned into. By using the risk data from their recent PHAs and HAZOPs, their own processes were now producing practical and understandable results and delivering them to the plant floor.

It's clear that for any initiative to be the best it can be, the participants must understand the perspective of others, especially those groups holding data and operating similar sites. Intention matters for this primary group of benefactors. They must witness demonstrations and pilots. They have chosen to use risk data as a safeguard, and they have chosen to do it without heavy lifting and spending.

The all-important task of helping other primary beneficiaries to change their thinking has already begun. Vendors are now expected to share all their information as data, for the benefit of the requester and the benefit of all. Vendors are now expected to leverage the data made available by the beneficiary to reduce their costs and increase their quality and efficiency, thereby benefitting all.

Who Owns the Problem?

It happened to us seven years ago, during an energy industry downturn that left us wondering what to do. In the future, it will happen to other groups as well. But for now, we are the ones delivering data as safeguards, and we are the ones demonstrating, piloting, and integrating the learnings for the beneficiary groups, both primary and secondary.

When the beneficiary group of owner-operators didn't see themselves as owning the problem, it was the risk professionals, the consultants, the established risk businesses, who quickly became the owners of the data as safeguards. By demonstrating and piloting data as safeguards, this group is slowly bringing together a once fragmented, siloed group of risk consultants and supporting them with people and tools. We are now a group of risk data technocrats, PHA analysts, and PHA facilitators, connecting to the world for all its risk consultants, and Uberizing the passion of all risk consultants.

What started out as a service to collect and ingest data before giving back the results has changed. The process started bringing people together and solving the "complete" risk picture. Not long after, predicting risk and unsafe days was possible.

This fairly new group includes techno savvy graduates, first and second career people, combined with seasoned risk professionals walking the internet as risk consultants. Now, with this team coming together in a cohesive way, the primary beneficiary can get exactly what they want: direct, industry-related process experience, and the intelligence that can be gleaned from the data. This quickly got the attention of everyone, even the old guard where traditional safeguards were so easy to rinse and repeat.

Who's Ready to Help Shape the Future?

I'll close with a message about looking to the future. If I have managed to convince you to consider a career in Risk & Safety, imagine a future where you are a change-maker. This career has the potential to make you happier and more excited; it will interest you and challenge you. Every day, you will encounter something new, making you stronger and adding more value to what you can offer your employer, your business, and your clients. You will be contributing to the world in new ways designed to solve an old problem, helping companies spend less, stay safe, and make the world a better place. You will be nothing less than a modern-day hero.

GLOSSARY OF TERMS

Term	Description	Metric
PSM	Process Safety Management involves implementing policies to prevent hazardous events.	Number of incidents or near misses prevented
PHA	Process Hazard Analysis uses techniques to identify and assess risks in industrial processes.	Number of hazards identified and mitigated
Risk Management	The systematic process of identifying, assessing, and controlling risks	Risk reduction percentage
Risk	The potential for loss or harm arising from a hazard or threat	Probability and impact assessment
Digital Risk	Risks associated with the use of digital technologies and data	Number of digital incidents or breaches
Unsafe day	A day during which a workplace experiences an incident or near miss	Number of unsafe days reported
Near miss	An event that could have resulted in harm but did not	Number of near misses reported
Data Safeguards	Measures to protect data from unauthorized access or breaches	Number of data breaches prevented

Term	Description	Metric
Risk Learnings	Insights gained from analyzing past risk events and near misses	Number of implemented improvements from learnings
Data Ingestion	The process of collecting and processing data for analysis	Volume of data ingested per unit time
Threat	A potential cause of an unwanted incident	Number of identified threats
Impact	The effect or severity of an incident on the organization	Severity rating of incidents
Safeguard	Measures or controls in place to prevent or mitigate hazards	Number of safe-guards implemented
Hazard Event	An occurrence that has the potential to cause harm	Frequency of hazard events
Enterprise	A business or organization involved in industrial activities	Overall risk profile score of the enterprise
Site Hazards	Specific risks or dangers present at a particular location or facility	Number of site-specific hazards identified

KEN BINGHAM has spent forty-plus years in Safety & Risk, dedicating thirty of those years to leading and innovating the industry with a digital risk consulting company that provides data as a service (DaaS), builds safety products, and provides risk services and training. Ken is an innovator, the CEO of RSD Group, and a past Ernst & Young Entrepreneur of the Year nominee. He is also an Engineering Technologist, a certified Electrical Journeyman, SIL/SIS Expert, TUV Instructor, and a past member of TEC Canada.

The Soul of Risk is Ken's second book, meant to inspire young graduates and mid-career professionals to consider a career in making the world safer with efficiency and profitability, and using a platform and the power of risk data. Ken has been happily married for fifty years, and is a proud father and grandfather, entrepreneur, rancher, and farmer.

www.ingramcontent.com/pod-product-compliance
Lightning Source LLC
Chambersburg PA
CBHW031853200326
41597CB00012B/395